# ENZYMES, ENERGY AND METABOLISM

## M.R.INGLE

B.Sc., Ph.D.

Head of Biology, Moreton Hall School,
Oswestry, Shropshire

BASIL BLACKWELL

Published by Basil Blackwell Ltd
108 Cowley Road
Oxford OX4 1JF
England

**British Library Cataloguing in Publication Data**

Ingle, M. R.
   Enzymes, energy and metabolism.—(Studies
   in advanced biology; 3)
   1. Biological chemistry     2. Metabolism
   I. Title     II. Series
   574.1'33     QP514.2

ISBN 0 631 90940 0 (*school edition*)
ISBN 0 631 14456 0 (*paperback*)

Typeset in 10/11½ pt Palatino
by Oxprint Ltd, Oxford

Printed in Great Britain

Other titles in this series:

*The Eukaryotic Cell: Structure and Function*
*Genetic Mechanisms*
*Microbes and Biotechnology*

# Contents

Most A-level students have access to a general text which provides a framework on which their course can be constructed. This series attempts to build on that framework by examining defined areas of the syllabus more closely. It looks especially at those parts of the subject which have recently undergone the most change, and attempts to bring together new concepts which are presently widely scattered throughout the available literature. In doing so, it tries to avoid simply substituting new dogmatic assertions for old, and to show how today's concepts have evolved from previous ones. A glossary is provided to avoid interrupting the text with too many asides.

This particular title posed the dilemma that a significant number of biology students may not have taken a formal course in chemistry and yet face examination questions which can be extremely taxing even to those who have (all the topics included have recently appeared in examinations in some form). I have tried to write the book making minimal assumptions about the reader's previous knowledge of chemistry, and to deal with specific issues when they arise. I have also attempted to write the book through a biologist's eyes rather than a biochemist's (at the risk of incurring the derision of both!), and so bridge the gap between the many excellent general biology and specifically biochemistry texts that are available.

MRI
1986

## Acknowledgements

I am most grateful to all those who assisted in the preparation of this book. In particular, I should like to thank Dr. Andrew Halestrap, Department of Biochemistry, University of Bristol Medical School, and Dr. David Hanke, Department of Botany, University of Cambridge, for their constructive comments on the manuscript. If any errors remain, it is not for want of trying on their part. I would also like to thank my wife, Rosemary, for typing the manuscript, and Bill Indge who drew the illustrations.

The author and publishers wish to thank the following for permission to reproduce material: Fig 3.1 Koch-Light Ltd.; Fig 4.5 Professor Ruth Bellaires, University College London; Fig 5.6 Science Photo Library; Fig 5.9 (2) Dr Steve Long, University of Essex; Fig 5.10 Dr Jean Whatley, Botany Department, University of Oxford; Fig 5.12 Biophoto Associates.

## Chemical nomenclature

The International Union of Physics and Chemistry (IUPAC) and International Union of Biochemists (IUB) have laid down precise rules for the naming of chemical compounds. Systematic names are used for those substances which the student is likely to meet on a concurrent chemistry course. However, many of the more unwieldy systematic names are not in common usage, and in these cases the trivial name is employed. Thus the text uses the term 'citric acid' rather than 2 hydroxypropane, 1.2.3 carboxylic acid, and 'lactose' rather than 4-0-(β-D-galactopyranosyl)-β-D-glucopyranose.

## Abbreviations

| | |
|---|---|
| $E$ | redox potential (V) |
| $F$ | Faraday's constant (96.5 kJ mole$^{-1}$ V$^{-1}$) |
| $\Delta G^{\circ}$ | Gibbs standard free energy change (kJ mole$^{-1}$) |
| $\Delta G$ | actual free energy change (kJ) |
| $h$ | Planck's constant ($6.626 \times 10^{-34}$ J s) |
| $\lambda$ | wavelength of light (nm) |
| ln | natural logarithm ($2.303 \log_{10}$) |
| M | moles per dm$^3$ |
| mole | the amount of a substrate which contains as many ions, atoms or molecules as there are atoms in 0.012 kg$^{12}$C |
| $N$ | Avogadro's number (the number of molecules, ions etc. in one mole: $6.023 \times 10^{23}$) |
| $R$ | universal gas constant ($8.31 \times 10^{-3}$ kJ mole$^{-1}$ K$^{-1}$) |
| $T$ | absolute temperature in Kelvin (K) |
| $z$ | number of electrons |

| | |
|---|---|
| C3 | an organic molecule containing three carbon atoms |
| C-3 | the *third* carbon atom in an organic molecule containing an unspecified number of carbon atoms |
| C3—COOH | an organic molecule containing three carbon atoms (one of which has a carboxyl group), i.e. total number of carbon atoms = 3 not 4 |
| citrate/citric acid<br>pyruvate/pyruvic acid | the terms for the dissociated and undissociated forms of organic acids are used interchangeably |

# Energy and Organisms

**SUMMARY**

The fundamental laws of physics and chemistry operate in living organisms and explain, at a molecular level, how organisms operate. Energy is needed to enable all chemical reactions to proceed. Organisms have developed mechanisms for storing chemical energy and using it to drive energy-consuming reactions.

The following terms are used:

aerobic respiration   anaerobic respiration   enzyme   glycolysis
hydrolysis   kilojoule (kJ)   mole   nucleotide   pH

The cell is a highly organised biochemical machine inside which most of the materials for an organism are processed, and most of an organism's energy is utilised and dissipated. In a single cell, scores and sometimes hundreds of reactions may be occurring simultaneously. In this book we shall consider why these reactions occur at all, why they happen at the speed they do, and how they are organised and regulated. In order to illustrate the discussion we shall examine those processes which contribute most substantially to the essential flow of energy through an organism.

## 1.1 THERMODYNAMICS

Organisms obey the laws of physics and chemistry. Indeed those characteristics which we call 'life' are not just consistent with such laws, but explained by them. Two laws of thermochemistry have outstanding significance in this context.

### The first law of thermodynamics

This law, which is also known as the **Principle of Conservation of Energy**, may already be familiar to the reader:

Energy cannot be created or destroyed, but it may be converted from one form into another.

By energy we mean the ability to do work – any kind of work – though in the context of this book we shall generally mean energy involved in converting one chemical substance into another.

From the first law it follows that one would have to express grave doubts if anyone claimed after a race to have 'used up a lot of energy'. They may have converted energy from one form to another, but they have not 'used it up' in the sense of destroying it.

**Q1** What form(s) would the energy take before and after a race?

**More on energy.** Energy can come in many forms, but here it will be heat, light and chemical energy which concern us most.

The last of these is especially important. The energy of a molecule is made up of several components. First, and most important, there is the energy of the electrons arising from their positions relative to the nucleus. In addition, there is the energy of the molecules as they move through space, rotate and vibrate. These forms of energy are intrinsic, i.e. they are an inherent property of matter.

### The second law of thermodynamics

This law may be stated in various ways, such as:

Processes involving the transformation of energy (i.e. work) will not take place spontaneously unless the total products of the transformation are more random than their precursors.

Our hypothetical athlete would be fully justified in saying that some *potential energy* (the capacity to do useful work, e.g. run) had been used up. During the race, the respiration of his/her blood sugar would not only have provided heat and chemical energy for muscle contraction, but would also have been converted into other forms. Among these other forms would be energy dissipated by increasing the purely random movement of molecules in and around the runner. Energy in this form, called **entropy**, is unusable.

As chemical reactions are a form of work, the laws of thermodynamics ought to have an important impact on the way in which we think about the many reactions occurring in a living organism. It follows from these laws that usable energy must be available for all chemical reactions to occur. Chemical reactions fall into two broad categories: those which occur of their own accord, without being supplied with external energy (**exergonic reactions**); and those which only occur if they are pushed along by an external energy supply (**endergonic reactions**).

## 1.2 EXERGONIC REACTIONS

Reactions in a living organism which involve breaking down complex organic molecules into simple ones (**catabolism**) are typically exergonic. The aerobic respiration of glucose is a classic example:

$$\underbrace{\text{oxygen} + \text{glucose}}_{\text{reactants}} \quad \overset{\text{form}}{\longrightarrow} \quad \underbrace{\text{carbon} + \text{water}}_{\text{products}} \atop \text{dioxide}$$

To accommodate the first law of thermodynamics we must assert that the left-hand side of the reaction exactly balances the right-hand side in terms of energy. Yet this reaction occurs without any outside energy being supplied, and the second law predicts that the products must somehow contain *less* intrinsic energy than the reactants. Are the laws contradicting each other? No, not really, because during the reaction some energy is released in various forms such as heat.

Chemists usually represent the difference between the energy content of reactants and products by the term $\Delta G^o$ (**Gibbs standard free energy change**). It is the energy difference between *one mole* of reactants and products at *equilibrium*, and is measured in kilojoules per mole (kJ mole$^{-1}$). By convention, if energy is released during a reaction it is given a negative sign. A complication arises if $H^+$ ions are involved, because $\Delta G^o$ would then be measured in a solution containing 1.0 mole $H^+$ per dm$^3$ solution. This solution would be pH 0! To allow for this the term $\Delta G^{o'}$ is normally used in biology. $\Delta G^{o'}$ is the same as $\Delta G^o$ but assumes pH 7: a value which approximates rather better to the aqueous conditions in the cell. A more adequate representation of aerobic respiration would therefore be

$$6O_2 + C_6H_{12}O_6 \rightarrow 6CO_2 + 6H_2O$$
$$(\Delta G^{o'} = -2870 \text{ kJ mole}^{-1})$$

$\Delta G^{o'}$ (or $\Delta G^o$) is helpful because it enables the energy changes of different reactions to be compared under a set of standard conditions. However, it has two very serious limitations:

(i) Reactants are never 1.0 M in living cells. They usually vary between 0.01 M ($10^{-2}$ M) and 0.00001 M ($10^{-5}$ M).
(ii) Reactions rarely reach equilibrium in living organisms. Frequently equilibrium is prevented by various cellular activities.

A knowledge of the actual free energy change, $\Delta G$, is therefore really more useful when trying to understand how organisms work. This takes into account the actual concentrations of materials, the pH, and does not assume equilibrium. It is, however, exceedingly difficult to measure.

## 1.3 ENDERGONIC REACTIONS

Reactions in the body involving the synthesis of complex molecules from simple ones (**anabolism**) are typically endergonic. Photosynthesis is a classic example:

$$6CO_2 + 6H_2O \rightarrow C_6H_{12}O_6 + 6O_2$$
$$(\Delta G^{o'} +2870 \text{ kJ mole}^{-1})$$

---

> **Q2** What information is provided in the equation which tells us that the above reaction cannot occur of its own accord?
>
> **Q3** Which law of thermodynamics tells us that the *energy* changes involved in aerobic respiration and photosynthesis are equal (but opposite)?
>
> **Q4** In photosynthesis, what form does the energy take (2870 kJ mole$^{-1}$) initially? Ultimately?

---

Endergonic reactions are driven in the body by coupling them to exergonic reactions, so that the energy change for the overall (coupled) reaction is favourable ($\Delta G^o$ or $\Delta G^{o'}$ becomes negative). Thus, suppose in an endergonic reaction:

$$A \rightarrow B \ (\Delta G^{o'} = +20 \text{ kJ mole}^{-1})$$

whilst in an exergonic reaction:

$$X \rightarrow Y \ (\Delta G^{o'} = -33 \text{ kJ mole}^{-1}).$$

If the two are coupled:

$$A+X \rightarrow B+Y \ (\Delta G^{o'} = (+20-33)= -13 \text{ kJ mole}^{-1}).$$

The coupling makes the overall $\Delta G^{o'}$ negative, so the reaction can proceed. The ultimate source of energy for the vast majority of endergonic reactions in organisms is light energy which is absorbed during photosynthesis. (The only exceptions are chemosynthetic bacteria, which substitute chemical energy derived from their environment for light.) These external forms of energy are used either directly or indirectly (via food which photosynthesis produces) to generate high energy intermediates. The latter are the *immediate* power supply for endergonic reactions.

Organisms are remarkably conservative in their choice of high energy intermediates, relying principally on just two main categories: **phosphorylated nucleotides** and **reduced dinucleotides**.

### 1.3.1 Phosphorylated nucleotides

The majority of endergonic reactions are powered by the phosphorylated nucleotide **adenosine triphosphate (ATP)**. Hydrolysis of the third phosphate group yields energy which can be made to do useful work ($\Delta G^{o'} = -30$ kJ mole$^{-1}$). Adenosine diphosphate (ADP) and inorganic phosphate (Pi) are byproducts of the reaction. Alternatively, ATP may be hydrolysed to adenosine monophosphate and pyrophosphate (AMP + PP) yielding a similar amount of energy. Hydrolysis of the remaining phosphate (AMP → adenosine + Pi) yields only a small amount of energy and is biologically insignificant (Fig. 1.1).

The first reaction is very common and is usually abbreviated to

$$ATP \rightarrow ADP + Pi + \text{energy}$$

as shown in Fig. 1.1 (inset), but this can be very misleading. Energy is never given out by breaking a

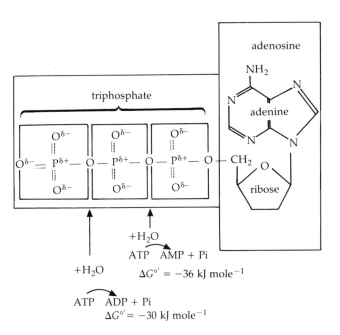

$+H_2O$

$ATP \quad AMP + Pi$

$\Delta G^{\circ'} = -36 \text{ kJ mole}^{-1}$

$+H_2O$

$ATP \quad ADP + Pi$

$\Delta G^{\circ'} = -30 \text{ kJ mole}^{-1}$

Energy is released when the last, or last two, phosphates are hydrolysed by an enzyme such as ATPase. Such enzymes normally require $Mg^{2+}$ in order to function. ATP is often loosely complexed to this ion in the cell.

$ATP^{4-} + H_2O \xrightarrow{ATPase} ADP^{3-} + Pi^{2-} + H_2O$
$(\Delta G^{\circ'} = -30 \text{ kJ mole}^{-1})$

---

glucose + phosphate → glucose phosphate
$(\Delta G^{\circ} = +16 \text{ kJ mole}^{-1})$

Since $\Delta G^{\circ}$ is positive (i.e. the reaction is energetically unfavourable), glucose phosphate cannot be synthesised by the above reaction.

$ATP \rightarrow ADP + Pi \quad (\Delta G^{\circ'} = -30 \text{ kJ mole}^{-1})$

However, the energy released during ATP hydrolysis is considerable, and can be used to drive the above, thus:

glucose + ATP → glucose phosphate + ADP
$(\Delta G^{\circ'} = -14 \text{ kJ mole}^{-1})$

Note that the energy released in the last reaction is the nett quantity left over from the first two reactions.

*Inset: Coupling of ATP to an endergonic reaction*

---

Fig. 1.1 *Adenosine triphosphate.* ATP is made from an organic base (adenine), a C5 sugar (ribose) and three phosphates. The significance of the small negative and positive charges ($\delta^-$ and $\delta^+$) is discussed at the end of Section 1.3.1. Various values for $\Delta G^{\circ'}$ are given in the literature, mostly between 30 and 33 kJ mole$^{-1}$.

chemical bond, as the abbreviation implies: in fact, energy is always required. The 30 kJ mole$^{-1}$ produced by ATP hydrolysis is the nett amount left over as a result of

| energy *released* during the formation of new bonds | minus | energy *required* to break existing bonds |

During respiration and photosynthesis, ATP is resynthesised from ADP and Pi. Thus the same molecules can be used over and over again; they can be made in one part of the cell and consumed in another.

Other phosphorylated nucleotides may be used to drive some endergonic reactions. Thus, whereas ATP is used for the synthesis of starch, protein and fat from their component parts, **guanidine triphosphate (GTP)** builds cellulose, and **uridine triphosphate (UTP)** builds glycogen, both from glucose. All the phosphorylated nucleotides yield a similar amount of energy when hydrolysed. Why ATP should be used so frequently as a power supply seems rather odd, and the energy content alone obviously cannot be the reason. Perhaps ATP is stable enough to prevent accidental wastage, but unstable enough to be broken down under appropriate conditions; perhaps enzymes using it can recognise it more easily.

---

**The squiggle and the energy in ATP**

The chemical energy potentially available in ATP is sometimes represented by allocating a 'squiggle' (= high energy bond) to the terminal phosphate, thus: ADP ~ Pi.

This has led to two misconceptions:

(i) *There is energy in the bond which is released when it is broken.* There is not. As stated above, energy is *always* required to break chemical bonds but if *more* energy is released when new (and more stable) bonds form as a result of a reaction then, on balance, energy will be released.

(ii) *There is something unique about this particular covalent bond.* There is not. It is likely, however, that in the cell the phosphate groups in ATP ionise, and therefore the molecule as a whole must contain more energy to maintain its integrity than in the absence of such charges. These charges are relieved on hydrolysis of the last (or last two) phosphate groups, resulting in the formation of more stable (lower energy) structures.

⟷ opposed charges which are relieved on hydrolysis

Hence it is incorrect to imagine that the energy in ATP is 'stored' in the terminal phosphate bond. For these reasons the term 'high energy bond' and the squiggle notation will not be used in this text.

### 1.3.2 Hydrogen (electron) carriers

Many biologically important reactions involve the gain or loss of electrons. These are called **redox reactions** (reduction–oxidation reactions). The substance which gains an electron is said to be reduced, and the substance which loses one is said to be oxidised. In cells the movement of an electron is usually accompanied by the movement of a proton (hydrogen ion) so that in most cases reduction means the gain of hydrogen, and oxidation the loss of hydrogen. Redox reactions in cells normally involve hydrogen (electron) carriers. Those which are encountered most often are **nicotinamide adenine dinucleotide (NAD), nicotinamide adenine dinucleotide phosphate (NADP)** and **flavin adenine dinucleotide (FAD)**.

$$\begin{array}{ccccc} \text{NAD} & + & 2\text{H} & \underset{\text{oxidation}}{\overset{\text{reduction}}{\rightleftharpoons}} & \text{NADH}_2 \\ \text{oxidised} & & & & \text{reduced} \\ \text{form} & & & & \text{form} \end{array}$$

The ability of chemicals to accept or donate electrons varies. Substance A might be able to donate electrons to substance B but not to substance C. Indeed C might donate electrons to A (and B). Hence we can arrange chemicals in a sequence according to their ability to donate electrons. In this example the sequence would be C → A → B. This ability may be quantified under standard conditions of concentration, temperature and pH. Electron-donating ability is described in terms of the **redox potential**, the units being volts. Table 1.1 lists some biologically important redox potentials. What the table means is that substances at the top can reduce (donate electrons or hydrogen to) those lower down. Thus $NADH_2$ can pass hydrogen to oxygen, forming water, but water cannot spontaneously pass hydrogen back to NAD to re-form $NADH_2$.

$$\begin{array}{ccccccc} \text{NADH}_2 & + & \tfrac{1}{2}\text{O}_2 & \longrightarrow & \text{NAD} & + & \text{H}_2\text{O} \\ \text{reduced} & & \text{oxidised} & & \text{oxidised} & & \text{reduced} \end{array}$$

From the laws of thermodynamics this must mean that the oxidation of $NADH_2$ by oxygen has a negative $\Delta G^{o'}$. Indeed, whenever an electron passes 'down' the table, $\Delta G^{o'}$ must be negative.

---

**Q5** Chemists have derived an equation which makes it possible to quantify the relationship between $\Delta G^{o'}$ of oxidation and redox potentials:

$$\Delta G^{o'} = - zFE$$

where $z$ = the number of electrons involved
  $F$ = Faraday's constant
    ($96.5$ kJ mole$^{-1}$V$^{-1}$) and
  $E$ = the redox potential.

The values of $\Delta G^{o'}$ given in the table were calculated using this equation. Can you work out the value for the oxidation of $NADH_2$ by oxygen? Proceed as follows:
(i) If electrons are passing from $NADH_2$ to $\tfrac{1}{2}O_2$, through how many volts in *total* (Table 1.1) do the electrons move? (What is $E$?)
(ii) From $\Delta G^{o'} = -zFE$, calculate $\Delta G^{o'}$ for this reaction. Remember that *two* electrons are involved, i.e. $z = 2$.

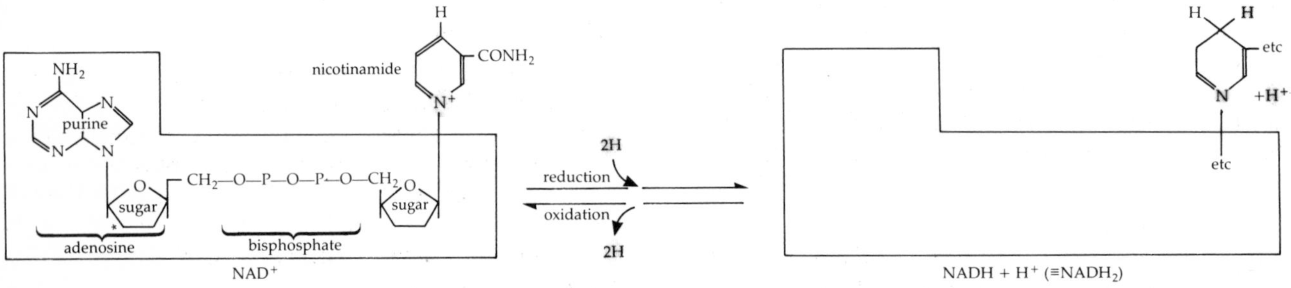

---

Ignore the detail in the left-hand drawing: it is only there to show how the component parts join together (adenine, ribose (twice), diphosphate and nicotinamide). Notice how NAD can pick up two hydrogens ( H ) and so function as a 'hydrogen' carrier. Strictly speaking, it is better to describe it as an *electron* carrier since, as the diagram indicates, N$^+$ is actually converted to N, not NH$^+$.

| Some electron (hydrogen) carriers | Full name | Comment |
|---|---|---|
| NAD | Nicotinamide adenine dinucleotide | Used in both anaerobic and aerobic respiration. In the latter, hydrogen is transferred from food materials to oxygen, forming water, and ATP is formed in the process (see Chemiosmosis, Chapter 4.4). NAD also substitutes for NADP during bacterial photosynthesis. |
| NADP | Nicotinamide adenine dinucleotide phosphate (same as NAD, with extra phosphate at position marked * in above figure) | Used in synthetic reactions such as photosynthesis and fatty acid synthesis where some of the intermediate compounds in these pathways are reduced. In photosynthesis, the ultimate source of the hydrogen is water, and its ultimate destination is organic acids, which are thus reduced to aldehydes (sugars). |
| FAD | Flavin adenine dinucleotide | Used at some stages of aerobic respiration in place of NAD. |

Fig. 1.2 *Common hydrogen carriers.* Nicotinamide is made from **nicotinic acid (niacin)**, one of the vitamin B group. Flavin is made from **riboflavin** (Vitamin B$_2$). The deficiency symptoms which appear when such substances are missing from the diet are essentially outward signs of inadequate quantities of NAD, NADP and FAD.

Table 1.1 *Redox potentials of some biologically important compounds.* The redox potential ($E'$), which is measured in volts, varies with pH. The values given assume biological conditions, i.e. pH 7.

| $E'$ (V) | Oxidised → Reduced | $\Delta G^{\circ\prime}$ of oxidation | |
| --- | --- | --- | --- |
| | | by $O_2$ | by $NO_3^-$ |
| $-0.43$ | $\left\{\begin{array}{ll}CO_2 & \to CH_2O \\ Fd_{ox} & \to Fd_{red}\end{array}\right.$ | $-239$ | $-164$ |
| $-0.32$ | $\left\{\begin{array}{ll}NADP & \to NADPH_2 \\ NAD & \to NADH_2\end{array}\right.$ | (See Q5 Chapter 1) | (See Q14 Chapter 4) |
| $-0.22$ | $FAD \to FADH_2$ | $-199$ | $-124$ |
| $-0.1$ | Range for most cytochromes and other components of cellular electron transport systems | | |
| $+0.38$ | $PSI^+ \to PSI$ | (Not biologically significant) | |
| $+0.42$ | $NO_3^- \to NO_2^-$ | $-75$ | — |
| $+0.81$ | $O_2 \to H_2O$ | — | — |
| $+0.9$ | $PSII^+ \to PSII$ | — | — |

Energy is released ($\Delta G$ is negative) if an electron moves from a (reduced) substance higher up the table to an (oxidised) substance lower down the table.

**Key**

Fd      ferredoxin (a chloroplast component: Chapter 5)
$CH_2O$    carbohydrate
PSI, PSII   photosystems I and II (Chapter 5)
Other abbreviations are defined in the text

The amount of energy which is released by the oxidation of reduced dinucleotides can, as your answer to Q5 shows, be considerable. It can be used in a variety of ways. The energy released during the oxidation of $NADH_2$ itself, for example, is used to build up ATP from ADP and Pi. Indeed it is by far the most important mechanism of ATP synthesis in most organisms. Similarly, during photosynthesis, $NADPH_2$ (which is built up by light energy from NADP and $H_2O$) provides about 80% of the energy needed to convert $CO_2$ into carbohydrate:

$$CO_2 \xrightarrow[\substack{2NADPH_2 \quad\quad 2NADP}]{\text{photosynthesis}} C(H_2O) + H_2O$$
$$\text{carbohydrate}$$

(ATP is also consumed in this reaction, and provides the remaining 20% of the energy required.)

### 1.3.3 Other high energy intermediates

The box in Section 1.3.1 emphasised that the energy content of ATP was a function of the electrical charges distributed throughout the phosphate groups. One might therefore expect to find that other phosphorylated compounds are also 'energy rich'. Table 1.2 confirms that this is indeed the case. Indeed, some micro-organisms even use polymers of *inorganic* phosphate as energy reserves.

Table 1.2 *Miscellaneous 'high energy' compounds.* $\Delta G^\circ$ for ATP hydrolysis is given in Fig. 1.1.

| Compound hydrolysed | $\Delta G^{\circ\prime}$ (kJ mole$^{-1}$) | Comments |
| --- | --- | --- |
| Phosphoenolpyruvate → pyruvate + phosphate | $-54$ | ATP-generating steps during anaerobic respiration |
| Diphosphoglyceric acid → phosphoglyceric acid + phosphate | $-51$ | |
| Creatine phosphate → creatine + phosphate | $-43$ | Energy-generating mechanism in vertebrate muscle |
| Acetyl coenzyme A → acetate + coenzyme A | $-31$ | Acetyl CoA is a source of organic carbon and energy for fatty acid synthesis, oxidative respiration and the synthesis of some amino acids |
| Sucrose → glucose + fructose | $-29$ | Energy-generating mechanism during inulin synthesis |
| Glucose-6-phosphate → glucose + phosphate | $-16$ | G6P is an activated form of glucose used in respiration |

The only biologically significant molecule in Table 1.2 which is not phosphorylated, but which nevertheless yields useful amounts of energy on hydrolysis, is **acetyl coenzyme A**. More will be said about this substance later.

### 1.3.4 Coupling in series

It was noted in Section 1.2 that in biology the term $\Delta G^{o'}$ had severe practical and theoretical limitations, and that $\Delta G$ was more useful. The two can be related thus:

$$\Delta G = \Delta G^{o'} + RT \ln \left( \frac{\text{product concentration}}{\text{substrate concentration}} \right)$$

where $R$ = Universal gas constant
$(8.31 \times 10^{-3} \text{ kJ mole}^{-1} \text{ K}^{-1})$
$T$ = temperature in Kelvin
$\ln = 2.303 \log_{10}$

Assuming organisms are at 25 °C, this equation simplifies down to

$$\Delta G = \Delta G^{o'} + 8.31 \times 10^{-3} \times 298 \times$$

$$2.3 \log_{10} \left( \frac{\text{product concentration}}{\text{substrate concentration}} \right)$$

$$\Delta G = \Delta G^{o'} + 5.7 \log_{10} \left( \frac{\text{product concentration}}{\text{substrate concentration}} \right)$$

Q6 shows why the equation is so useful.

---

**Q6** Suppose that for a given reaction $\Delta G^{o'} = +8 \text{ kJ mole}^{-1}$, but that in the conditions found in the cell the substrate concentration is 0.04 M, and the product concentration is 0.001 M. Calculate $\Delta G$, and state whether or not, under these conditions, the reaction can proceed.

---

The equation shows that a reaction can be made possible (i.e. have a negative $\Delta G$) if the substrate concentration is kept very high, and the product concentration is kept very low. This is possible if a highly exergonic reaction either precedes or follows an endergonic one: the two become **sequentially coupled**, or linked 'in series'. It is a relatively common phenomenon. About half the steps in anaerobic respiration are endergonic, but these are coupled in series to others which are strongly exergonic, and for the pathway as a whole $\Delta G^{o'} = -65 \text{ kJ mole}^{-1}$.

## 1.4 THERMODYNAMICS AND LIFE

W. B. Yeats, the Irish poet, put the second law of thermodynamics in a nutshell. 'Things', he said, 'fall apart'. Continents drift, galaxies explode and dead organisms decay. There is nothing special in these examples; they are simply different expressions of how the amount of disorder (entropy) in the universe inevitably tends to increase. Let us imagine a very simple universe indeed, one consisting entirely of a metal rod, hot at one end and cold at the other. Heat will gradually move from the hot to the cold end, and during this time the transfer of heat could, in principle, be made to do useful work. Eventually, however, the heat will spread throughout the rod until the temperature is uniform, after which no further change will take place. Our model universe has come to an end of its life. In a similar way the 'real' universe is, according to the second law of thermodynamics, irrevocably 'running down'. Energy is *not* being lost (first law) but useful (potential) energy is being converted into non-useful forms. Entropy is increasing.

Living organisms are a peculiar localised reversal of the natural order. This reversal is expressed in organised growth, in the molecular architecture of tissues and cells, and in the synthesis of complex organic molecules. It is true that relatively disorganising events occur as well, but on balance entropy is *decreasing* during life. This localised reversal of the general trend towards increasing disorder is only possible at the expense of the rest of the universe. The maintenance of biological order – homeostasis in its broadest sense – requires a constant external supply of materials and energy.

Materials can be used and discarded, to be re-used by other organisms, so resulting in the endless recycling of carbon, nitrogen and other elements. In sharp contrast there is a one-way flow of energy through the living world, beginning with photosynthesis and ending, for example, in its dissipation as heat by carnivorous mammals. Organisms are rather like a huge, complex, endergonic reaction, in which $\Delta G$ can only be made favourable by coupling them to an external energy supply (light). It is this external supply of energy which results, in the case of every living organism, in a localised reversal of the second law of thermodynamics.

---

**Study guide**

*Vocabulary*

Distinguish between the following pairs of terms:

endergonic and exergonic reactions
reduction and oxidation
$\Delta G^{o'}$ and $\Delta G$

*Review Question*

Relate the structure and properties of ATP to its function in living organisms. Illustrate your answer by reference to its involvement in *one* of the following:
(i) muscle contraction
(ii) the sodium pump
(iii) photosynthesis
(iv) starch synthesis.

*Extension Question*

Why does a biologist need to know about the second law of thermodynamics?

---

# Enzymes

## SUMMARY

Enzymes are protein catalysts which increase the rates of chemical reactions to the speeds needed to sustain life. They work by reducing the amount of energy which reactant molecules must possess before they can undergo a chemical change. Their unique properties are exploited by organisms for the regulation of metabolism.

The text itself defines several new terms, but the following are assumed (see Glossary if required):

equilibrium    product    substrate    isomer
reactant    X-ray crystallography    pH

An appreciation of energetics tells us why some chemicals *can* react together and others cannot, but nothing whatever about how *fast* a reaction will proceed. The questions 'Is it possible?' and 'How quickly?' are quite different. The fact that the oxidation of cellulose is thermodynamically extremely favourable, for example, does not mean that this book will immediately burst into flames. To understand what governs the speed of a reaction we must turn to **enzymes** and a concept called **activation energy**.

## 2.1 CHARACTERISTICS OF ENZYMES

In 1897 the Buchner brothers showed that sugar could be fermented by the juices extracted from ground-up yeast cells. They called the active juice '**enzyme**' ('in yeast'; Gk.). Soon other extracts with the same ability to speed up chemical reactions were discovered. The term 'enzyme' was given to all these, and a new name, *zymase*, was given to the mixture of substances isolated by the Buchners. The chemical nature of enzymes was uncertain until Sumner (1926) first purified and crystallised the enzyme *urease*. It was a protein. Over 250 enzymes have since been purified, and without exception they have been proteins. The vast majority of enzymes are **intracellular** (operate inside cells). The more familiar digestive enzymes of the gut are therefore rather atypical, since these are all **extracellular**.

A **catalyst** is a substance which speeds up a chemical reaction. Enzymes are, by definition, organic catalysts. They resemble inorganic catalysts, such as platinum, in several important respects. For example:

(i) they remain unaltered at the end of the reaction;
(ii) they are not used up during the reaction;
(iii) small quantities are extremely effective;
(iv) neither alters the end products of the reaction;
(v) neither alters the equilibrium position of a reaction. The amount of product at the *end* of a reaction is the same whether or not a catalyst is present.

However, enzymes differ from inorganic catalysts in several important respects. Thus, enzymes are:

(i) affected by pH;
(ii) **denatured** (destroyed) by high temperatures;
(iii) affected by the presence of other substances such as coenzymes, cofactors and inhibitors;
(iv) highly specific, only catalysing one reaction, or one type of reaction. An enzyme which catalyses a single reaction is said to show **absolute specificity**. An enzyme which attacks one type of chemical bond in a variety of substrates, e.g. a peptide bond, is said to show **group specificity**.

These unique differences are all related to the fact that enzymes are proteins. The relationship between an enzyme and a substrate has been likened to that between a **lock** and a **key** (Section 2.2.2). In other words, the *shape* of an enzyme molecule is critically important for its normal functioning. The structure and shape depend on the various covalent and electrical bonds holding the molecule together (Fig. 2.1). It follows that factors which affect these bonds will affect the shape of the enzyme, and thus its activity. Heat, cofactors and pH ($H^+$ ions) are examples of such factors. The last of these also affects electrical charges on various ionised amino acids in the enzyme molecule. As we shall see (Section 2.2.3), having the right electrical charge in the right place at the right time is crucial for an enzyme's catalytic activity.

### 2.1.1 Types of enzymes

Over 90% of enzymes are simple **globular proteins**, the remainder being **conjugated proteins** with non-protein **prosthetic groups** (Fig. 2.1). For biological purposes, however, classifying enzymes by their physical characteristics is much less useful than classifying them by their function. The Enzyme Commission of the International Union of Biochemists (1964) opted for the latter approach, and listed six basic categories (Table 2.1).

Each of the six main categories is assigned a number: thus all hydrolases are in group 3. Each group is then subdivided several times and eventually the enzymes are individually numbered. The complete classification

**Generalised amino acid** **1**

non-ionised     both —NH₂ and —COOH ionised

$$NH_2 - C_\alpha - COOH \qquad NH_3^+ - C_\alpha - COO^-$$

(with H above and R below each $C_\alpha$)

The degree of ionization on the amino (—NH₂) and carboxyl (—COOH) groups varies with pH.

About 20 types of amino acids occur in proteins. They differ from each other only by the variable side chain (R). Some of the broad categories of amino acids, together with specific examples, are:

**Hydrophilic and acidic**

Glutamate
$R = —(CH_2)_2COOH$

**Sulphur containing**

Cysteine
$R = —CH_2SH$

**Hydrophilic and basic**

Lysine
$R = —(CH_2)_4NH_2$

**Hydrophobic**

Phenylalanine
$R = —CH_2$ —⬡

---

Amino acids combine by the formation of peptide bonds to form long chains called **polypeptides**. The number, type and arrangement of amino acids in a polypeptide depends on the protein, and is called the **primary structure** of the protein. **2**

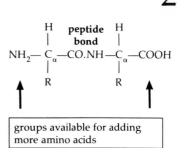

$$NH_2 - C_\alpha - CO.NH - C_\alpha - COOH$$
(peptide bond; with H above and R below each $C_\alpha$)

| groups available for adding more amino acids |
| --- |

---

Between 20% and 80% of the polypeptide chain normally organises itself into configurations called α-helices or β-pleated sheets. These are called the **secondary structures** of the protein. **3**

α-helix: more elastic

hydrogen bonds

In an α-helix, hydrogen bonds form between the —NH₂ group on the α-carbon of one amino acid, and the —COOH group of the α-carbon four amino acids further along the polypeptide.

β-pleated sheet: greater tensile strength

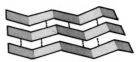

hydrogen bonds between polypeptides

In a β-pleated sheet, adjacent polypeptides are joined by hydrogen bonding between neighbouring peptide bonds. Some fibrous proteins, e.g. collagen, never form α-helices or β-pleated sheets.

---

Most of the remaining polypeptide now folds in a manner dictated by the properties of the constituent amino acids, to form the **tertiary structure**. The precise shape of the three-dimensional molecule which is formed plays a significant role in determining the properties and hence the function of the protein. **4**

The tertiary structure is maintained by a few disulphide bridges (—S—S—) formed between sulphydryl (—SH)-containing amino acids, and a variety of weak electrical bonds such as hydrogen, dipole and electrostatic bonds. The latter are strongly affected by H⁺ ions, and readily disrupted by temperature. Hence pH and heat markedly affect protein activity. Hydrophobic bonds are also present. These create small water-free zones in the protein.

---

The final product may take various forms. **6**

*Simple globular proteins:* highly folded polypeptide chain(s). Over 90% of proteins are in this class.

*Simple fibrous proteins:* linear polypeptide chains. Keratin and collagen are in this class.

*Conjugated proteins:* mostly globular proteins, all combined to a non-protein component. If the compound is a tightly bound organic compound, it is called a **prosthetic group** (haem, of haemoglobin, being an example). If it is a small detachable molecule or ion it is usually called a **cofactor**.

---

Some proteins consist of a single polypeptide. In others, two or more polypeptides combine to form the **quaternary structure**. In many cases, such as haemoglobin, the constituent polypeptides are not identical. **5**

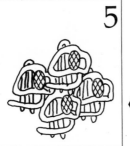

---

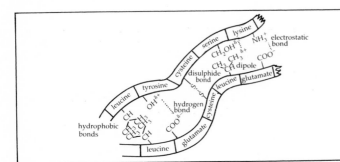

Fig. 2.1 *The structure of proteins*

Table 2.1 *Classification of Enzymes (IUB system; 1964)*

| Number | Enzyme category | Type of reaction | Examples |
|---|---|---|---|
| 1 | Oxidoreductases | Oxidation/reduction reactions. Two types: oxidases and dehydrogenases | *Oxidases:* transfer hydrogen to oxygen, e.g. cytochrome oxidase |

$$Cyt\ H_2 + \tfrac{1}{2}O_2 \rightleftharpoons Cyt + H_2O$$

reduced cytochrome      oxidised cytochrome

*Dehydrogenases:* transfer hydrogen to a molecule other than oxygen

e.g. *lactate dehydrogenase*

$$C_3H_6O_3 + NAD \rightleftharpoons C_3H_4O_3 + NADH_2$$

lactate      pyruvate

This reaction occurs in liver when the oxygen debt is paid off following heavy exercise.

e.g. *glutamate dehydrogenase*

$$glutamate + H_2O \longrightarrow \alpha\text{-ketoglutarate} + NH_3$$

NAD    NADH$_2$

This reaction is an important mechanism for destroying excess amino acids. It occurs in liver cells, and is called *deamination*.

| 2 | Transferases | Transfers a functionally important group from one molecule to another | *Transaminases:* transfer amino groups, making new amino acids from existing ones |

e.g. *aspartate transaminase*

aspartate + α-ketoglutarate $\rightleftharpoons$
(amino acid$_1$)   (carboxylic acid$_2$)

glutamate + oxaloacetate
(amino acid$_2$)   (carboxylic acid$_1$)

Occurs in all cells.

*Kinases:* transfer phosphate from (usually) ATP to another substance

e.g. *hexokinase*

$$ATP + glucose \longrightarrow ADP + glucose\text{-}6\text{-}phosphate$$

This reaction 'activates' glucose prior to its breakdown in respiration.

*Phosphorylases* (add inorganic phosphate without using ATP) are also in this category.

| 3 | Hydrolases | Split molecules in two by the action of water | All digestive enzymes fall into this category: pepsin, trypsin etc |

e.g. *amylase*

$$starch + H_2O \rightarrow starch + maltose$$

(one disaccharide   (disaccharide)
shorter)

These reactions occur in the gut and transform large insoluble food molecules into smaller soluble ones which can be absorbed. Similar reactions also occur in the lysosomes of cells.

*Phosphatases* (remove phosphate groups from organic molecules by hydrolysis) are also in group 3.

*(continued)*

Table 2.1 *Classification of Enzymes (IUB system; 1964) (continued)*

| Number | Enzyme category | Type of reaction | Examples |
|---|---|---|---|
| 4 | Lyases | Add or remove groups without involving water | *Carboxylases*: add $CO_2$ <br><br> e.g. *ribulose bisphosphate carboxylase* <br><br> ribulose bisphosphate + $CO_2$ → 2 phosphoglyceric acid <br> C5 sugar $\qquad\qquad$ 2 × C3 organic acid <br><br> This occurs in the stroma of chloroplasts and is the principal mechanism by which atmospheric $CO_2$ is turned into organic material. <br><br> *Decarboxylases*: remove $CO_2$ <br><br> e.g. *pyruvate decarboxylase* <br><br> pyruvic acid → ethaldehyde + $CO_2$ <br> C3 $\qquad$ C2 $\qquad$ released <br><br> Intermediate step in the conversion of sugars to ethanol during alcoholic fermentation. |
| 5 | Isomerases | Convert one isomer (form) of a compound into a different isomer by redistributing the atoms | *Mutases* <br><br> e.g. *phosphoglucomutase* <br><br> glucose-1-phosphate $\rightleftharpoons$ glucose-6-phosphate <br><br> This reaction occurs early in the respiration of sugars. |
| 6 | Ligases | Link together two molecules at the expense of ATP | *Synthetases* <br><br> e.g. *aminoacyl synthetases* <br> Join amino acids to tRNA during protein synthesis. |

of α-*amylase*, for example, is 3.2.1.1. Table 2.1 is more than just a catalogue. It implies that all the biochemicals in a cell can be formed from six basic types of reaction, which is a considerable achievement considering the enormous diversity of substances present.

Most enzymes end in '-ase', as in α-*amylase*, above. The first part of the name often indicates the substrate on which the enzyme is acting: *amyl*ase (enzyme) acts on *amyl*ose (starch). Although the names of enzymes can be extraordinarily long, they are at least descriptive. Hence '*DNA-dependent RNA-polymerase*' indicates that this enzyme (-*ase*) makes RNA by polymerisation of nucleotides provided that DNA is present. Unfortunately there are some exceptions to these rules. For example, many digestive enzymes are known by their historical names, such as pepsin, trypsin etc. There is no alternative but to memorise these.

---

Look at Table 2.1.
**Q1** Name the types of enzymes which:
(i) add $CO_2$
(ii) convert one isomer into another
(iii) remove hydrogen
(iv) break up large molecules into small molecules using water
(v) make new amino acids by transferring amino groups from existing ones
(vi) 'energise' molecules by adding Pi from ATP.
You will meet these enzymes several times in the following chapters.

---

## 2.2 THE MECHANISM OF ENZYME ACTION

### 2.2.1 Activation energy

If a reaction *can* take place (is exergonic) then it *will* take place, but it may do so very slowly. Paper does burn in air at room temperature, but extremely slowly. The brittle yellow pages of old manuscripts testify to a slow rate of combustion.

The rate of reaction is determined by how many molecules have the **activation energy** ($E_A$) to react together at any one moment. In a slow reaction only a small percentage of the molecules involved have the necessary amount. The activation energy therefore represents a kind of barrier which must first be overcome before the reaction can proceed (Fig. 2.2).

In principle, reactions can be speeded up in two ways:
(i) by supplying the reactants with more energy, for example by heating them;
(ii) by lowering the activation energy by means of a catalyst.

In organisms, there are limits to the amount of heat that can be supplied, because heat progressively denatures proteins. Among the proteins which could be damaged are, of course, the enzymes (Section 2.3.2). Enzymes lower activation energy barriers very dramatically. Without enzymes in your gut, it would take about 50 years instead of a few hours for your last meal to be hydrolysed (digested). Before we consider how enzymes lower the activation energy, it is necessary to take a closer look at the lock and key hypothesis.

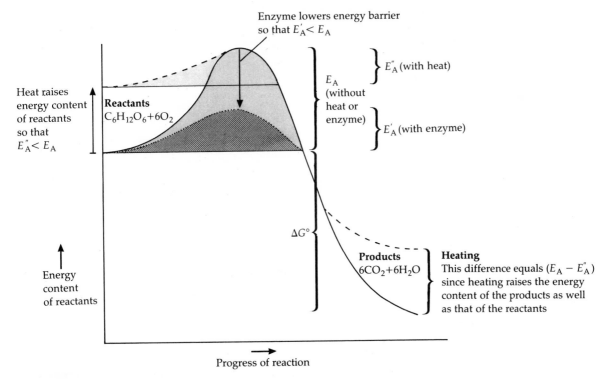

Fig. 2.2 *Activation energy: heat and enzyme action.* A secondary effect of heat is that it increases the *kinetic energy* of the reactants, so improving the chances of a collision. Although this contributes to an increase in rate, the effect is small.

## 2.2.2 The lock and key hypothesis

Emil Fischer's lock and key hypothesis (1894) has been an immensely useful concept. Fischer originally developed the idea in order to explain enzyme specificity. This it does admirably. For a key to work it must be provided with the right lock, and so it is with enzymes and substrates. Like all useful hypotheses, it makes testable predictions. In fact, it makes two.

### The ES complex

The first prediction is that if enzymes and substrates really are analogous to locks and keys, then some

reaction between an enzyme and a substrate must occur, however brief. We can therefore represent a reaction thus:

$$E + S \rightleftharpoons ES \rightleftharpoons E + P$$
enzyme substrate      complex      enzyme product

The most direct evidence for the formation of transient ES complexes is that they can actually be isolated from enzymes that work rather slowly. For example, under appropriate conditions chymotrypsin forms relatively stable chymotrypsin–protein complexes when the substrate (protein) is added. Other evidence comes from a technique called spectroscopy. Look at Fig. 2.3 and then Q2.

> **Q2** A spectrophotometer is an instrument which measures the wavelengths of light absorbed by a substance. These wavelengths vary for different chemicals, and each chemical has a distinctive **absorption spectrum** (see Appendix). In the case of an enzyme, the wavelength it absorbs changes when a substrate is added, indicating a change in the enzyme. Thus in Fig. 2.3, graph A becomes graph B when equimolar concentrations of *catalase* and hydrogen peroxide are mixed. A moment later, graph B disappears and graph A reappears. What conclusions may be drawn from these results? (There are about four.)

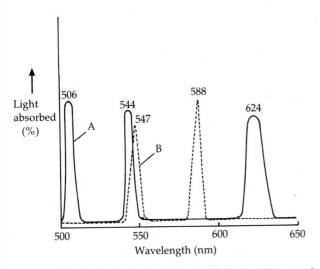

Fig. 2.3 *Spectroscopic evidence for an ES complex.* The graph shows the absorption spectrum for *catalase* before (—) and after (---) adding the substrate ($H_2O_2$).

### The concept of the active site

A second prediction is that there must be one or more **active sites** on an enzyme which are the centres of catalytic activity. Here we face a dilemma: enzymes are enormous, with a molecular mass of tens or hundreds of thousands, but substrates are often small. So are these (predicted) sites scattered all over the enzyme, or are there, say, just one or two at a fixed position? See if you can draw your own conclusions from Q3 and Q4.

---

**Q3** *The number of substrate molecules which bind to an enzyme at any one moment in time.*
(i)   What is suggested by the fact that *equimolar* concentrations of substrate and enzyme produce the changes shown in Fig. 2.3?
(ii)  With a different enzyme and substrate, rather different results were obtained (Fig. 2.4). In this experiment, the wavelength of light absorbed by the substrate was measured at various concentrations of enzyme. What conclusions may be drawn from this graph?

**Q4** *The position at which substrate molecules bind to an enzyme.*
A *substrate analogue* resembles a true substrate but binds very tightly to an enzyme, forming a stable ES complex. In one experiment an analogue always inactivated chymotrypsin when it bound *to one particular* site on the latter.
(i)   What conclusion may be drawn from this experiment?
(ii)  What assumption are you making in drawing this conclusion?

---

Experiments like those described in Q3 and Q4 support the view that substrate molecules bind only one or two at a time to specific points on an enzyme. They are consistent with the idea, if not direct proof, that enzymes possess specific active sites for catalysis, as implied by Fischer's lock and key hypothesis. **X-ray crystallography** is another powerful tool for determining the three-dimensional structure of an enzyme and its active site. Figure 2.5 shows the results of one such analysis on the enzyme *lactate dehydrogenase*.

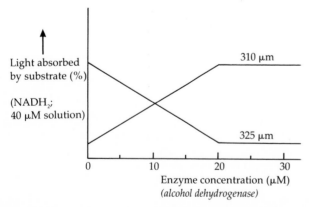

Fig. 2.4 *To determine the number of substrate molecules which bind to an enzyme at any one time: spectroscopic analysis of substrate (NADH$_2$) at increasing levels of enzyme (alcohol dehydrogenase)*

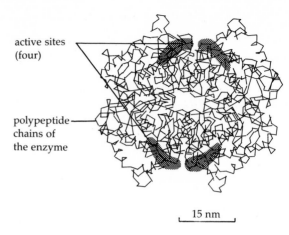

Fig. 2.5 *The structure of lactate dehydrogenase*

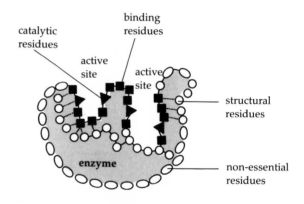

Fig. 2.6 *Amino acid residues in an enzyme*

If the active site is only a small part of the enzyme molecule, what is the rest of the molecule doing? Koshland (1963) suggested that an enzyme consists of essentially four categories of amino acids (Fig. 2.6):
(i)   *Catalytic residues (catalytic site)*
      These make and break chemical bonds. They are the basis of catalytic activity.
(ii)  *Binding residues (binding site)*
      These hold the substrate in place while catalysis is taking place.
The catalytic and binding residues together form the active site.
(iii) *Structural residues*
      These hold the active site in the correct shape so that it can function properly.
(iv)  *Non-essential residues*
      These have no specific function. They are often near the surface of an enzyme and can be removed or replaced without loss of function.

*Locks and keys reconsidered: the theory of induced fit*

Evidence from protein chemistry suggests that a slight rearrangement of chemical groups occurs in both enzyme and substrate when an ES complex is formed. Enzymes are therefore best regarded as rather flexible molecules whose shape can change slightly under the influence of electrical charges present on the substrate during the formation of a complex. This idea, of an enzyme 'wrapping round' a substrate to form a more stable structure, is called the **induced-fit hypothesis**.

## 2.2.3 The molecular basis of enzyme action

In thermodynamic terms, enzymes work by lowering the activation energy barrier by re-routing the overall reaction through several steps, each with much smaller activation energies (Fig. 2.7). These small steps include the transient formation of an ES complex as described previously.

In molecular terms, several mechanisms probably operate simultaneously at the active site, all of which contribute to a lowering of the activation energy and hence to an increase in the rate of a chemical reaction (Table 2.2).

Table 2.2 *Molecular mechanisms which contribute to a lowering of the activation energy*

| Method | Comment |
|---|---|
| 1 Proximity effects | Temporary binding of reactants next to each other on an enzyme increases the chance of a reaction |
| 2 Strain effects | Slight distortion of the reactants as they bind to the enzyme strains the bonds which are to be broken and increases the chances of breakage |
| 3 Orientation effects | Reactants are held by the enzyme in such a way that bonds are exposed to attack |
| 4 Micro-environmental effects | Hydrophobic amino acids create a water-free zone in which non-polar reactants may react more easily |
| 5 Acid–base catalysis | Acidic and basic amino acids in the enzyme facilitate the transfer of electrons to and from the reactants |

The last item in Table 2.2 warrants special attention. Figure 2.8 illustrates how it could work for the hydrolysis of a peptide bond. Thus:

(i) Electrons on catalytic site $X^-$ pull protons away from the O atom in water. The O atom therefore becomes more negatively charged.

(ii) Protons on catalytic site $Y^+$ pull oxygen away from the C atom on an amino acid. The C atom consequently becomes more positively charged.

(iii) As a result of the increased charge on the water's *oxygen* (i) and the amino acid's *carbon* (ii) these components tend to react together, and the peptide bond is hydrolysed.

The effectiveness of acid–base catalysis depends on the precise position of positive and negative electrical charges on the amino acids of an enzyme. These charges will be affected by the pH of the solution (Section 2.3.1). Since pH has a dramatic effect on enzyme activity we may conclude that acid–base catalysis is a very important catalytic mechanism.

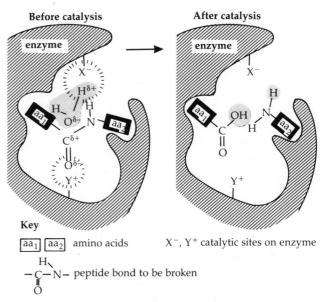

Fig. 2.8 *Acid–base catalysis*

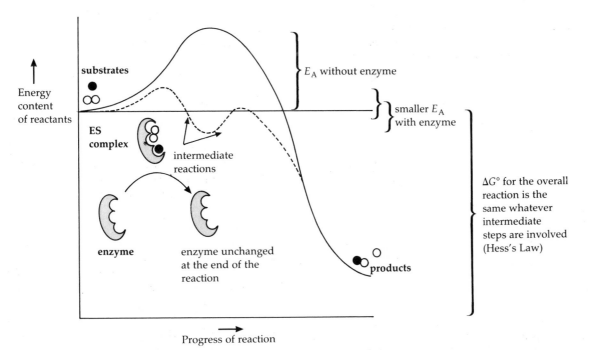

Fig. 2.7 *The re-routing of a reaction by an enzyme*

## 2.3 FACTORS AFFECTING THE RATE OF ENZYMIC REACTIONS

The rate of an enzymically controlled reaction can be influenced by several factors.

> **Q5** Three factors which affect enzyme activity (and hence the rate of reaction) have already been mentioned. What are they?

In addition to these factors (Q5) the rate is also affected by:
(i)   enzyme concentration;
(ii)  substrate concentration;
(iii) product concentration.

When the effects of one factor are experimentally investigated under *in vitro* (test-tube) conditions, all others must be kept constant as far as possible. This is not always easy, particularly in the case of substrate and product concentration. Variation in these can be minimised, however, by using very large amounts of substrate and running the experiments for very short times (minutes, and even seconds).

> **Q6** Explain how the procedures mentioned in the last sentence minimise variation in substrate and product concentration.

*In vitro* experiments undoubtedly contribute significantly to an understanding of enzymic reactions, but it is emphasised that in the organism itself the factor which is most critical may vary from time to time and place to place.

### 2.3.1 pH

All enzymes function most effectively at a characteristic pH (the **optimum pH**), which is usually pH 7 ± 1.5. Outside this range, the activity of an enzyme decreases, and finally stops (Fig. 2.9). *Pepsin* (a proteolytic enzyme) has an unusually low optimum. This is an adaptation to the acid conditions of the stomach where it operates (stomach pH 1.5–2.5 due to 0.1%–0.5% HCl in the gastric juice).

> **Q7** If an enzyme is subjected to a pH slightly different from its optimum, its activity will decrease. However, full activity can be restored if the pH is shifted back to the optimum. If an enzyme is subjected to a pH very different from its optimum, then its activity will be irreversibly diminished no matter how it is subsequently treated. Explain these observations.

### 2.3.2 Temperature

The effect of temperature on thermochemical reactions is normally expressed mathematically by the **temperature coefficient** ($Q_{10}$). This tells us how a 10 °C rise in temperature affects the rate of reaction:

$$Q_{10} = \frac{\text{rate of reaction at } (x + 10) \text{ °C}}{\text{rate of reaction at } x \text{ °C}}$$

In living organisms the rate of reaction generally doubles for a 10 °C rise ($Q_{10} = 2$) within the normal range of environmental temperatures.

The effects of temperature on enzymically controlled reactions are complex. A rise in temperature increases both the rate of reaction and also the rate of destruction of the enzyme catalysing the reaction (enzyme destruction is itself a chemical reaction). Most enzymes are rapidly denatured above about 50 °C due to loss of quaternary and tertiary structure. Consequently, in a living organism the **optimum temperature** will be a compromise between the effect of heat on the reaction itself, the rate at which an enzyme is denatured and the rate at which new enzymes can be made by the cell. Mammals and birds, which maintain high constant body temperatures (**endotherms**), have an optimum of about 35–40 °C. Plants and other animals (**ectotherms**) generally have lower optimum temperatures, between 20 °C and 35 °C. Some prokaryotes are exceptional and are adapted to life in hot springs at over 80 °C.

Although organisms can be said to have optimum temperatures, enzymes themselves most certainly cannot. *In vitro* experiments may well be performed resulting in graphs such as Fig. 2.10, though the

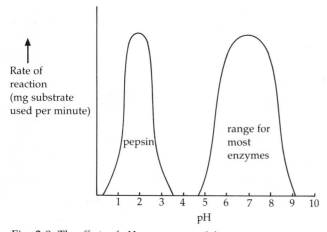

Fig. 2.9 *The effects of pH on enzyme activity*

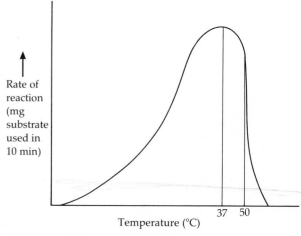

Fig. 2.10 *Effects of temperature on the rate of enzymically controlled reaction*

only useful comment which can be made about them is that they are skewed to the left, indicating rapid **denaturation** (destruction) above about 50 °C. The 'peak' is not an optimum. If the experiment had been done over 10 seconds instead of 10 minutes, the peak might well have been at 60 °C or 70 °C.

**Q8** Identical mixtures of enzyme and substrate were incubated at 30 °C and 50 °C. Samples were taken after 30 seconds and 10 minutes, and the product concentrations were measured. From the results below, calculate the mass (µg) of product formed per minute for each of the four samples. Comment on the significance of the figures you have calculated.

|  | 30 s sample (µg product) | 10 min sample (µg product) |
|---|---|---|
| 30 °C | 2.5 | 46.5 |
| 50 °C | 8 | 10 |

### 2.3.3 Enzyme concentration

The rate of an enzymically controlled reaction is directly proportional to the enzyme concentration under constant conditions (Fig. 2.11). Clearly one way of starting or stopping a particular chemical reaction in a cell might be to make or destroy a particular enzyme. As it turns out, this is only one of several regulatory methods, and in fact is a rather crude one (Chapter 3).

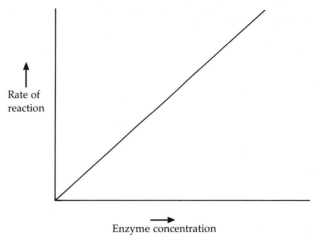

Fig. 2.11 *The effect of enzyme concentration on the rate of reaction*

### 2.3.4 Substrate concentration

The presence of an enzyme does not only speed up a reaction, it also has a marked effect on its kinetics (course). Without an enzyme, the rate of reaction is directly proportional to the substrate concentration [Fig. 2.12(i)]. In the presence of an enzyme, two kinds of responses may occur [Fig. 2.12(ii); Fig. 2.12(iii)]. These last two responses are similar in that eventually the rates level off, so that increasing substrate concentration eventually has no further effect. They differ in that with some enzymes the reaction 'takes off' immediately [Fig. 2.12(ii)], whereas in others it 'takes off' more slowly [Fig. 2.12(iii)].

Figure 2.12(ii) is known mathematically as a rectangular hyperbola, and an enzyme producing such a graph is said to show **hyperbolic** or **Michaelis–Menten**

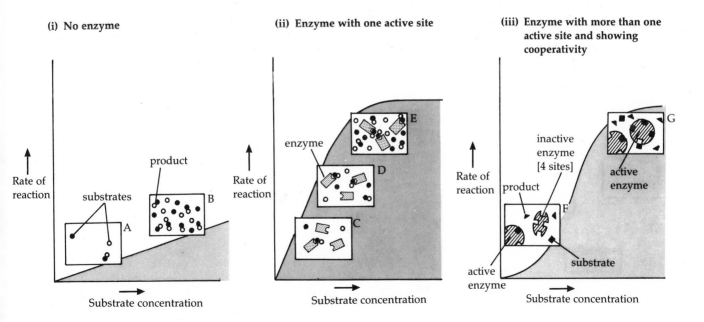

Fig. 2.12 *Reaction kinetics*

**kinetics**. Figure 2.12(iii) is described as sigmoidal (S-shaped), and an enzyme producing it is said to show **sigmoidal kinetics**. The differences reflect some fundamental features of enzymes.

*Hyperbolic kinetics*

Not much product is formed per minute at low substrate concentrations, simply because there is not much substrate from which it can form (region C). As the substrate concentration rises (region D), the active site on the enzyme has more material to work on, so the rate of reaction increases. At high substrate concentrations, the active sites are so saturated with substrate that the product cannot be made any faster (region D). Hyperbolic kinetics are typical of, but not exclusive to, enzymes composed of a single polypeptide chain with one active site. An example is *trypsin*.

*Sigmoidal kinetics*

Enzymes composed of more than one polypeptide and with more than one active site sometimes exhibit sigmoidal kinetics. S-shaped curves are generally attributed to **cooperativity** or cooperative effects. This means that, as the substrate concentration rises (region F), binding of some substrate molecules alters the shape of the enzyme in such a way that other substrate molecules bind more easily. At very low substrate concentrations, cooperativity is less, so the reaction is slow to 'take off'. The rest of the graph (region G) is similar to that for hyperbolic kinetics.

The binding of oxygen to haemoglobin (four polypeptides, each with one oxygen-binding site) produces a sigmoidal curve. The similarity between the temporary binding of oxygen to haemoglobin and the formation of an ES complex has prompted some biochemists to call haemoglobin an 'honorary enzyme'.

### 2.3.5 Inhibitors

Substances which reduce enzyme activity fall into two main categories:

(i) *Irreversible inhibitors*

These bind tightly to enzymes, altering them so that they permanently lose their catalytic properties. Arsenic, lead, mercury and various insecticides are all irreversible inhibitors. Even at low concentrations the effect of heavy metals can be so drastic that enzymes are actually precipitated.

(ii) *Reversible inhibitors*

These bind less tightly to enzymes, and their inhibiting effects can be reversed. There are two main types:

(a) *Competitive inhibitors (CIs)*

These do not affect the catalytic ability of an enzyme, but because they resemble the substrate molecule they compete with the latter for the active site. As a result, the ability of the substrate to bind to the enzyme is reduced. This means that a higher substrate concentration than normal is needed to achieve the same rate of reaction. $CO_2$, for example, is the normal substrate for the photosynthetically important enzyme *ribulose bisphosphate carboxylase*. This

enzyme is competitively inhibited by $O_2$, and even relatively low $O_2$ concentrations reduce the rate at which $CO_2$ is incorporated into sugars (Chapter 5.5).

(b) *Non-competitive inhibitors (NCIs)*

These bear no resemblance to the normal substrate, but bind to the enzyme in such a way that they reduce its catalytic properties. Consequently even though the substrate will bind to the enzyme in the normal way, the latter works less effectively and the rate of reaction will remain reduced until the inhibitor concentration drops. It follows from this that, in contrast to competitive inhibition, simply adding more substrate will not restore the rate of reaction to its previous value.

---

**Q9** Figure 2.Q9 shows the results of an experiment under three sets of conditions:
(i)   enzyme + substrate
(ii)  enzyme + substrate + CI
(iii) enzyme + substrate + NCI.
State which is which, giving a reason.

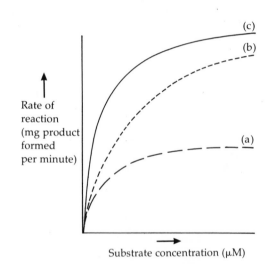

Fig. 2.Q9 *The effects of competitive and non-competitive inhibitors on the rate of reaction (see Q9)*

**Q10** Dialysis tubing allows small molecules to pass through, but not enzymes. How, in principle, could you use this material to distinguish a NCI from an irreversible inhibitor?

---

Competitive and non-competitive inhibitors are not necessarily harmful. Quite the reverse: both are used extensively by cells for metabolic regulation (Chapter 3).

### 2.3.6 Cofactors, coenzymes and prosthetic groups

Some substances actively promote enzyme activity. Three main groups are recognised whose similarities are listed in Table 2.3. Any attempt to draw clear distinctions between these accessory substances runs into the problem that similar substances may behave differently in different circumstances. Moreover, different authorities tend to use the same terms in

Table 2.3 *Similarities between cofactors, coenzymes and prosthetic groups*

1 They all help enzymes (or other proteins) to function, but they have no catalytic properties of their own

2 They are not themselves proteins (not composed of amino acids) and are invariably much smaller than the enzymes with which they are associated

3 They may undergo a temporary change during a reaction, but they are normally restored afterwards

4 They are all required in small amounts

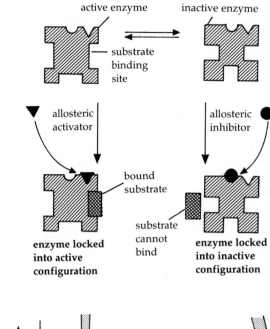

different ways. Some biologists, for example, would apply the term 'cofactor' to *all* types of substance which promote enzyme activity. Bearing these difficulties in mind, we shall distinguish them here as follows.

**Cofactors** are mostly ions (especially divalent cations) or small molecules which normally bind reversibly to an enzyme and can thus easily be removed by dialysis. Some promote enzyme activity because they are components of the active site, but many are not. To understand how the latter work it is necessary to divert our attention for a moment to the concept of allosteric regulation.

Table 2.4 *Examples of cofactors*

| Cofactor | Enzyme | Comment |
|----------|--------|---------|
| $Ca^{2+}$ | Phosphorylase kinase | Speeds glycogen breakdown: allosteric activator. (See cAMP: Section 3.2.3.) |
| $Zn^{2+}$ | Carboxy-peptidase A | Component of active site: in Fig. 2.8 'Y' = $Zn^{2+}$ |
| $Mg^{2+}$ | ATPase | $Mg^{2+}$ is often bound to ATP in the cell |
| $Cl^-$ | Salivary amylase | Allosteric activator |
| AMP | Phospho-fructokinase | Allosteric regulator of glycolysis [Chapter 4] |

**Allosteric regulators** are substances which inhibit or activate enzymes but which do not resemble the reactants, and though they bind reversibly to enzymes they never bind to the active site. To explain how they exert their effects it is necessary to assume that enzymes can exist in 'active' or 'inactive' configurations (shapes). Accordingly, **allosteric activators** lock enzymes into an active configuration, whilst **allosteric inhibitors** lock them into an inactive configuration (Fig. 2.13). Complex situations may arise where a substance is an allosteric activator of one enzyme and an allosteric inhibitor of another. Given such diversity, it is perhaps not surprising to find that allosteric regulators play a significant role in metabolic regulation. The kinds of substance which can act as allosteric regulators are equally diverse. Some are complex organic molecules, whereas others are inorganic ions.

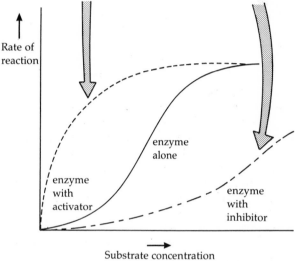

Fig. 2.13 *Allosteric regulators and their effects on enzyme kinetics. Phosphofructokinase,* a key regulatory enzyme in sugar respiration, is an example of an allosteric enzyme. It is activated by AMP and inhibited by ATP.

**Coenzymes** are larger, organic molecules which bind reversibly and transiently to an enzyme. In many respects they are more akin to substrate molecules, but unlike substrate molecules they are regenerated and can therefore, like the enzyme itself, be re-used. Examples include NAD, NADP, FAD and coenzyme A (CoA). The first three have already been mentioned as electron carriers (Fig. 1.3). The last converts low energy carboxyl groups (R.COOH) to high energy acyl groups (R.CO. *SCoA*) as in **acetyl CoA** (Table 1.2).

**Prosthetic groups** are non-protein groups tightly bound to proteins. It is not normally possible to remove them without disrupting the rest of the molecule. One category of substances which commonly form prosthetic groups is the **porphyrins**. These contain a metal ion, such as $Fe^{2+}$, at the centre (Fig. 2.14). Iron-containing porphyrins are called **haem**. The enzyme *catalase* contains haem, as do several important non-enzymic proteins such as the **cytochromes** (Chapters 4.4 and 5.2.1) and haemoglobin itself.

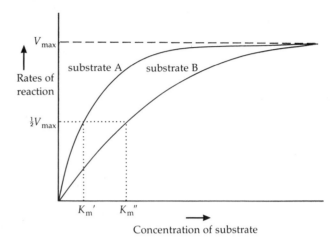

Fig. 2.14 *Porphyrins as prosthetic groups.* Ignore the details: concentrate on the overall shape of the basic porphyrin ring and note the metal ion ('X') held in the centre.

## 2.4 QUANTITATIVE TREATMENT OF ENZYME ACTION

> Students should consult their tutors as to whether a quantitative treatment of enzyme kinetics is required in the syllabus which they are following.

The majority of enzymes show hyperbolic kinetics, and this account applies specifically to these. It is convenient to begin by defining two terms shown in Fig. 2.15, $V_{max}$ and $K_m$.

### $V_{max}$

This is the maximum velocity (rate of reaction) attainable. The higher $V_{max}$ is, the more product is formed per minute, so it is essentially a measure of the catalytic effectiveness of the enzyme. The units are usually given as $\mu$moles of product formed per minute (or moles per second [katals]). It is a useful parameter, because if the concentration of enzyme is known the **turnover number** can be calculated:

$$\text{turnover number} = \frac{\text{moles of product formed per minute}}{\text{moles of enzyme}}$$

Turnover numbers are rarely less than $10^4$. The highest is for *carbonic anhydrase*, each molecule of which converts about $3.6 \times 10^7$ $CO_2$ molecules to $H_2CO_3$ per minute:

$$H_2O + CO_2 \xrightarrow[\text{(in erythrocytes)}]{\text{carbonic anhydrase}} H_2CO_3$$

### $K_m$ (Michaelis constant)

This is a constant unique to a particular enzyme for a particular substrate. It is defined as the concentration of substrate necessary to produce half the maximum velocity. It is typically of the order of $10^{-3}$ to $10^{-6}$ $\mu$M substrate concentration. A *small* $K_m$ means that only a *small* substrate concentration is needed to attain maximum velocity (compare $K_m'$ and $K_m''$, Fig. 2.15). It is an indication of the affinity of an enzyme for a particular substrate: the *smaller* the $K_m$, the *greater* the affinity.

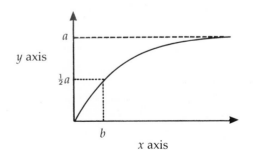

Fig. 2.15 *The implications of $V_{max}$ and $K_m$.* Although $V_{max}$ for the enzyme is the same for both substrates, $K_m'$ is much smaller than $K_m''$, so $V_{max}$ is reached sooner with substrate A than with substrate B.

> **Q11** By knowing the $K_m$ it should be possible to predict which of two possible substrates a given enzyme will attack. Under what conditions would the enzyme in Fig. 2.15 attack (i) both substrates, (ii) substrate A preferentially.

Michaelis and Menten first showed that graphs like Fig. 2.15 conform to what are known mathematically as rectangular hyperbolas, the general expression for which is

$$y = \frac{ax}{x + b}$$

where $a$ = maximum value of $y$ reached ($\equiv V_{max}$)
$b$ = value of $x$ when $y = \frac{1}{2}a$ ($\equiv K_m$)

and in our case
$x$ axis $\equiv [S]$ (substrate concentration in $\mu$M)
$y$ axis $\equiv v$ (velocity of reaction in $\mu$moles product min$^{-1}$)

**Q12** Substitute the symbols $V$, $[S]$, $V_{max}$ and $K_m$ into the above equation. The result you obtain is called the Michaelis–Menten equation.

The determination of $K_m$ and $V_{max}$ from curves such as those shown in Fig. 2.15 is difficult, since in a hyperbola the graph only levels out completely at infinity. However, Lineweaver and Burk showed that this problem is easily overcome by a double reciprocal plot, i.e. $1/V$ against $1/[S]$. This turns $y = ax/(x + b)$ into a straight line, so that $V_{max}$ and $K_m$ can be calculated directly (Fig. 2.16).

*Inhibitors*

Competitive and non-competitive inhibitors have easily distinguishable effects on $V_{max}$ and $K_m$. With competitive inhibitors (which do not alter the active site) the $V_{max}$ is not affected. This is because $V_{max}$ is defined by an *infinite* amount of substrate, at which point the inhibitor concentration must be zero. However, the ability of the enzyme to bind with a substrate is reduced by competition, so that $K_m$ *increases* (Fig. 2.17). Conversely, since non-competitive inhibitors bind to the enzyme in such a way as to reduce its catalytic properties, $V_{max}$ will be permanently reduced until the NCI is removed. However, in this case the ability of the enzyme to bind with the substrate is not affected, so that $K_m$ remains unaltered (Fig. 2.17). The extension question in the study guide may help to illustrate these rather complicated arguments.

*Sigmoidal kinetics*

It is emphasised that enzymes showing sigmoidal kinetics do not conform to the above treatment. The mathematics involved are more complex and fall outside the scope of this book.

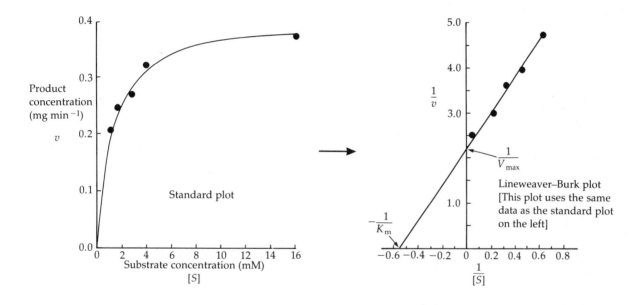

For reference only:

Given $v = \dfrac{V_{max}[S]}{K_m + [S]}$

Then by taking reciprocals we obtain:

$\dfrac{1}{v} = \dfrac{K_m + [S]}{V_{max}[S]}$

And by expanding the right-hand term we obtain:

$\dfrac{1}{v} = \dfrac{K_m}{V_{max}[S]} + \dfrac{[S]}{V_{max}[S]}$

The last item, $\dfrac{[S]}{V_{max}[S]}$, can be simplified by dividing through by $[S]$, giving us:

$\dfrac{1}{v} = \dfrac{K_m}{V_{max}[S]} + \dfrac{1}{V_{max}}$ (i)

Now the equation for a straight line is $y = mx + c$
And equation (i) fits this where:

$\dfrac{1}{v} = y;\quad \dfrac{K_m}{V_{max}} = m;\quad \dfrac{1}{[S]} = x;$ and $\dfrac{1}{V_{max}} = c$

Now if $x = 0$, then $y = c$,

i.e. if $\dfrac{1}{[S]} = 0$, then $\dfrac{1}{v} = \dfrac{1}{V_{max}}$

Also, if $y = 0$, then $x = \dfrac{-c}{m}$,

i.e. when $\dfrac{1}{v} = 0$, then $\dfrac{1}{[S]} = \dfrac{1}{-K_m}$.

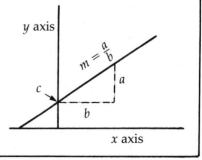

Fig. 2.16 *The Lineweaver–Burk plot for determining $V_{max}$ and $K_m$.* For practical purposes, only the plot need be known. The mathematical explanation of why the intercepts correspond to $-1/K_m$ and $1/V_{max}$ is included for reference purposes only.

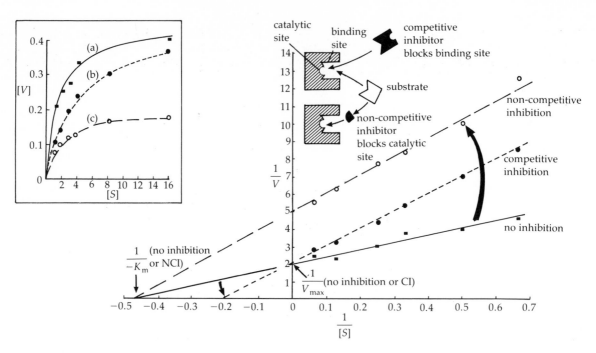

Fig. 2.17 *Lineweaver–Burk plots showing the differences between competitive and non-competitive inhibitors for an enzyme showing hyperbolic (Michaelis–Menten) kinetics.* The inset shows the same data, but not plotted as reciprocals.

## Study guide

*Vocabulary*

Distinguish between:
(i)  peptide, hydrogen, disulphide and hydrophobic bonds;
(ii)  allosteric regulators, coenzymes, cofactors and prosthetic groups;
(iii)  $E_A$ and $\Delta G^o$;
(iv)  irreversible, competitive and non-competitive inhibitors.

*Review Questions*

1  What properties do enzymes share with inorganic catalysts? What properties are unique to enzymes, and what is the biological significance of these unique differences?

2  (i)  How do enzymes work?
   (ii)  The amino (—NH$_2$) group of amino acids can be removed by the enzyme *glutamate dehydrogenase* which is normally present in mitochondria. Deamination (—NH$_2$ removal) is an essential preliminary step when amino acids are being used as an energy source during respiration. What conclusions can be drawn about the characteristics of this enzyme from the following data?

| Substrate concentration (mM) | 1.5 | 2.0 | 3.0 | 4.0 | 10.0 | 16.0 |
|---|---|---|---|---|---|---|
| Product concentration (mg min$^{-1}$) | 0.21 | 0.25 | 0.28 | 0.33 | 0.44 | 0.4 |

3  (Many *vitamins* are used for the synthesis of coenzymes. Use the following question to link the work covered in this chapter with the related topic of nutrition.)
What is a *vitamin*? List the characteristics of vitamins and by reference to named examples describe their principal functions in living organisms.

*Extension Question*

In an experiment the activity of samples of 10 μg of enzyme (molecular mass = $10^{-5}$ g mole$^{-1}$) each in 1 cm$^3$ of solution was tested at various substrate concentrations and in the presence of two substances, A and B. The results are shown below.

| [S] (mM) | 1 | 2 | 5 | 10 | 20 |
|---|---|---|---|---|---|
| Product concentration (μmoles min$^{-1}$) with A present | 0.77 | 1.25 | 2.00 | 2.50 | 2.86 |
| with B present | 1.17 | 2.10 | 4.00 | 5.70 | 7.70 |
| neither A nor B | 1.5 | 4.0 | 6.3 | 7.6 | 9.0 |

(i)  On the same pair of axes, graph these results in an appropriate way.
(ii)  Calculate $V_{max}$ and $K_m$ for the enzyme, in the presence of substrate alone.
(iii)  Calculate the turnover number for this enzyme in molecules of substrate per minute per molecule of enzyme.
(iv)  What conclusions can be drawn about substances A and B?
How do you explain their effects?

# 3

# *The Regulation of Metabolism*

**SUMMARY**

Since most important biological reactions only take place at a significant rate
in the presence of active enzymes, the most effective method of regulating
metabolism is by controlling the activity, quantity, location or kind of
enzymes present in a cell
The following terms are assumed:
covalent bond   ionic bond   translation   endoplasmic reticulum
plasma membrane   transcription   gene activation
An outline knowledge of the mechanism of protein synthesis
would be useful.

Figure 3.1 outlines some of the major metabolic pathways which can operate simultaneously in living cells. Somehow this tangled mass of reactions must be controlled. The cell must regulate the reactions of each individual pathway, and the overall rates of different pathways. An understanding of control mechanisms is clearly essential if we are to appreciate how cells and organisms function. These mechanisms, though complex at the molecular level, can be reduced to a few fundamental principles.

## 3.1 FUNDAMENTAL PRINCIPLES

Since most biologically important reactions only occur at a significant rate in the presence of enzymes, enzymes are the key to metabolic control. Regulation is achieved by altering enzyme activity, quantity and location, and the type of enzyme produced.

Regulation depends largely on one fundamental property of enzymes: specificity. If an enzyme is absent or inactive, no other can take its place, and a reaction will stop. Yet at the same time unrelated reactions will not be affected.

## 3.2 ENZYME ACTIVITY

The alteration of enzyme activity is undoubtedly the most important method of metabolic regulation on a minute-by-minute basis. In Chapter 2.1 the factors which affect enzyme activity were listed. Among these we can largely ignore pH and temperature so far as metabolic control is concerned. Their effects are so general that they tend to switch all enzymes 'on' or 'off' simultaneously (some use being made of this occasionally, e.g. during hibernation). However, inhibitors (Chapter 2.3.5) and accessory substances (Chapter 2.3.6) can be extremely specific and affect just one enzyme. Hence, by utilising these the cell has a system for controlling enzyme activity with great precision, and therefore for controlling individual reactions and pathways.

Table 3.1 *Methods of metabolic regulation.* Which method of control is used in a particular instance will vary. Each has different implications, and natural selection ensures that the most appropriate is used. After studying Sections 3.2 and 3.3, you should be able to work out which method is most appropriate in a particular set of circumstances.

| Method | Comments |
|---|---|
| 1 Regulating enzyme activity | Widely used for regulating minute-by-minute activities. Most suitable where a rapid response to changing conditions is required. Several methods of regulation are known: by the reactants themselves, by the end products, by secondary messengers, and by covalent modification (Section 3.2) |
| 2 Regulating the quantity of enzyme present | Achieved mainly by controlling the rate of enzyme synthesis, either at the level of **transcription** or at the level of **translation** (Section 3.3.1). The rate of enzyme degradation may also vary |
| 3 Regulating the location of enzymes | Compartmentalisation of related enzymes makes for more efficient organisation of the cell (Section 3.4) |
| 4 Regulating the type of enzyme produced | Different cells have the capacity to produce different enzymes, affecting different reactions. In addition, different cells may produce subtly different forms of the *same* enzyme (isozymes) with slightly different characteristics. Isozymes make possible small but biologically significant differences between tissues |

### 3.2.1 Feedback regulation

**Negative feedback** means that the product of a reaction or pathway inhibits its own production. For example, A is converted to B, and an accumulation of B inhibits

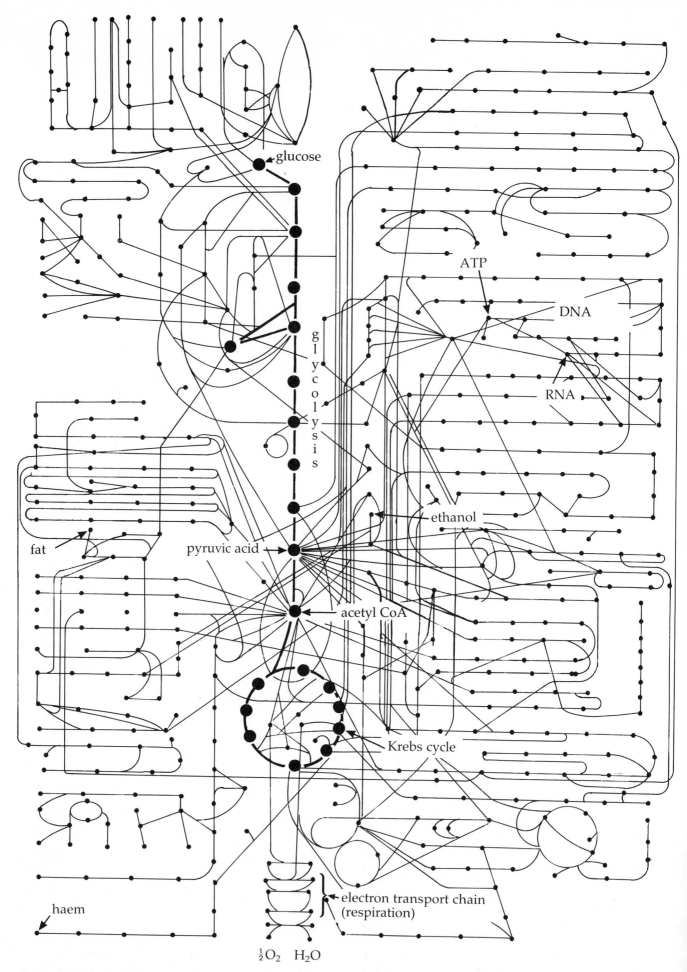

Fig. 3.1 *Major metabolic pathways*

the enzyme producing it. This is called **product inhibition**. As the concentration of B falls, the enzyme becomes more active again.

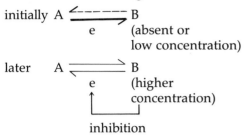

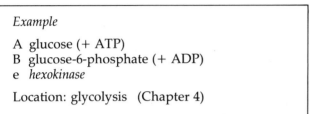

inhibition

---

*Example*

A glucose (+ ATP)
B glucose-6-phosphate (+ ADP)
e *hexokinase*

Location: glycolysis  (Chapter 4)

---

Often, the metabolites at the end of a pathway regulate an enzyme early in the pathway, so that the whole pathway is affected. The enzyme which is regulated is called the **pacemaker**. End-product regulation of an *early* step in a pathway is highly desirable. First, the accumulation of reactive and potentially hazardous intermediates is avoided. Secondly, because all reactions dissipate some energy into useless forms (increase entropy), the more unnecessary reactions that are shut off, the less energy is wasted. A specific example of a pacemaker system is described in Section 4.1.1.

### 3.2.2 Covalent modification

Some enzymes are activated by **covalent modification**. Often this involves the addition of a phosphate group:

$$e \; + \; ATP \longrightarrow e{-}P \; + \; ADP$$

inactive             active
enzyme             enzyme

---

*Example*

e *glycogen phosphorylase*
Location: first step in glycogen breakdown
(See Fig. 3.3)

---

Covalent bonds are strong, and as a result covalent modifications tend to be fairly stable. Indeed, sometimes they are irreversible, as in the case of the digestive enzymes. If active forms of trypsin, pepsin and other digestive enzymes were synthesised by the pancreas and alimentary canal, autodigestion of the cells producing them would be the inevitable and catastrophic result. They are therefore produced in inactive forms called **zymogens** (syn: **proenzymes**), and only when they reach the gut lumen do conditions exist for their covalent modification and activation.

The zymogens illustrate that metabolic control extends beyond individual cells. However, such irreversible systems are too crude to be much use in most circumstances. A more sophisticated alternative involves what are known as secondary messengers.

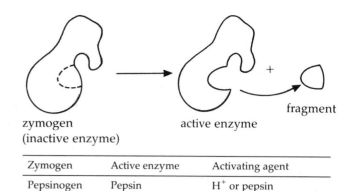

zymogen             active enzyme         fragment
(inactive enzyme)

| Zymogen | Active enzyme | Activating agent |
|---|---|---|
| Pepsinogen | Pepsin | $H^+$ or pepsin |
| Trypsinogen | Trypsin | Enterokinase or trypsin |

Fig. 3.2 *Conversion of zymogens to active enzymes*

### 3.2.3 Secondary messengers

Communication between one part of the organism and another enables it to function as an integrated whole. All integrating systems, whether nervous or hormonal, ultimately exert their effects through the regulation of events at the molecular level.

Some hormones, such as the sex hormones, are lipid derivatives. These are able to penetrate plasma membranes and seem to exert their effect mainly through gene activation (Section 3.3.1). Other hormones, such as adrenalin, ADH and glucagon, are modified amino acids or small proteins. These, the **primary messengers**, are unable to penetrate the plasma membrane. They do, however, bind to its outer surface, and in doing so increase the concentration of another substance, the **secondary messenger**, on the cytoplasmic side. The latter then diffuses into the cytosol, where it alters enzyme activity and so brings about responses characteristic of the hormone.

Only two secondary messengers are known for certain: **cyclic adenosine monophosphate (cAMP)** and $Ca^{2+}$. The former seems to be restricted to animal cells and bacteria. The latter seems to act as a secondary messenger in all living organisms.

*cAMP*

The cAMP system involves the following steps:

(i) *A hormone binds to and activates a receptor protein on the plasma membrane*
The binding is specific because the receptors vary with the cell type. Hence adrenalin binds to the plasma membrane of skeletal muscle cells but not to gut epithelial cells. Consequently, hormones will affect one tissue but not another.
(ii) *The hormone-bound receptor activates adenylate cyclase, which results in cAMP production*
The activation is probably via an intermediate (G-protein). *Adenylate cyclase* converts ATP to the secondary messenger cAMP (Fig. 3.3).
(iii) *Cytoplasmic cAMP triggers off a chain reaction among enzymes in the cytosol*

The action of every molecule of cAMP is greatly amplified in the cytosol by what is called an **enzyme cascade**. The system which generates glucose phosphate from glycogen is best understood. First, cAMP

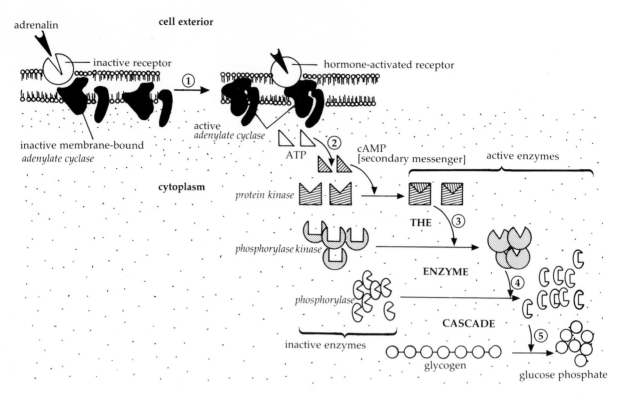

**Fig. 3.3** *cAMP production and the enzyme cascade.* The essence of the system is to produce as big a response as possible to a single adrenalin molecule. This is achieved by amplifying the original signal millions of times. At **1**, a single hormone molecule activates a membrane-bound protein. At **2**, each of these produces many 'secondary messengers' (cAMP) per second. cAMP then activates an enzyme, each molecule of which activates thousands of other enzyme molecules, **3**. This is repeated, **4**, until finally an enzyme is activated which breaks down stored food into glucose phosphate, **5**. The illustration applies to animals only. There is no evidence for the existence of equivalent systems in plants.

binds to and activates *protein kinase* (Fig. 3.4). Each activated *protein kinase* then catalyses the activation of many *phosphorylase kinase* molecules. Each of the latter catalyses the activation of many *phosphorylase* molecules. Finally, each *phosphorylase* molecule catalyses the breakdown of glycogen to glucose phosphate.

---

**Q1** Distinguish between (i) a kinase and (ii) a phosphorylase (Table 2.1).

**Q2** The action of cAMP perhaps seems unnecessarily complicated, but it is actually extremely effective. Suppose, for argument, that one hormone molecule *did* activate one phosphorylase molecule, and that the latter had a turnover number of $10^3$. How many substrate molecules would be produced in 5 s? Far more activated *phosphorylase* is produced by the enzyme cascade. Thus, suppose that one hormone molecule produces $10^3$ cAMPs during the first second; each cAMP then activates a single *protein kinase* in the next second; each of these activates $10^4$ *phosphorylase kinases* in the third second; each of these activates $10^4$ *phosphorylases* in the fourth second; and each of these produces $10^4$ glucose-6-phosphate molecules in the fifth second. How many glucose-6-phosphate molecules would be produced by the cascade system?

---

The cAMP-producing complex in the membrane rapidly breaks down. Moreover, hormones and cAMP are both rapidly destroyed by enzymes. As a result, hormonal effects do not persist indefinitely.

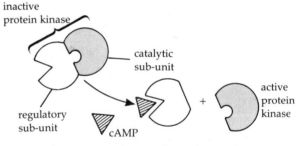

**Fig. 3.4** *The activation of protein kinase by cAMP*

### $Ca^{2+}$

The $Ca^{2+}$ level in the cytosol is usually very low ($10^{-7}$M) compared with that in interstitial fluid or in organelles such as sarcoplasmic reticulum (endoplasmic reticulum of muscle cells). Disturbance of the plasma membrane by a nerve impulse alters its permeability so that $Ca^{2+}$ floods into the cytosol causing the concentration to rise by up to $10^3 \times$. $Ca^{2+}$ now binds to a cytosol protein called **calmodulin**, and the complex acts as an allosteric activator of several important enzyme systems. It can, for example, activate *phosphorylase kinase*, reinforcing the activity of cAMP if the latter is present.

The high $Ca^{2+}$ level in the cytosol is, like cAMP, very transient. After an impulse has passed, the membrane rapidly reverts to its normal $Ca^{2+}$ impermeable condition, and $Ca^{2+}$ is actively pumped out of the cytosol at the expense of ATP by membrane-bound $Ca^{2+}$ pumps.

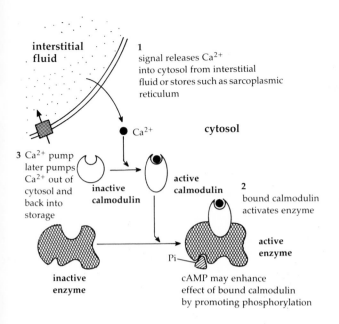

Fig. 3.5 *The action of $Ca^{2+}$ as a secondary messenger*

## 3.3 ENZYME QUANTITY

Although this topic is treated more fully in *Genetic Mechanisms*, Chapter 6, in this series, a brief review with a slightly different emphasis is appropriate here.

The amount of enzyme present depends upon at least two factors:
(i) the rate of synthesis;
(ii) the rate of destruction.
In the case of extracellular enzymes the amount present will also depend upon factors which control their release from the cell. Secretin, for example, controls the release of pancreatic juice. (This hormone was, incidentally, the first to be purified.)

### 3.3.1 Synthesis

It is well established that DNA determines the precise sequence of amino acids in proteins, and that protein synthesis involves **transcribing** (copying) the genetic code into an intermediary (mRNA), which is subsequently **translated** (decoded) into proteins. Regulation of protein synthesis can occur at the level of transcription (prior to mRNA formation) or translation (prior to decoding mRNA).

$$DNA \xrightarrow{\text{transcription}} mRNA \xrightarrow{\text{translation}} \text{protein (enzyme)}$$

*Transcriptional control*

Enzymes associated with the regulation of key metabolic pathways, such as respiratory enzymes, always tend to be present. They are described as **constitutive**. Other enzymes are produced in response to particular conditions, such as the presence of an appropriate substrate. These are called **inducible enzymes**. Working with bacteria, Jacob and Monod (1961) showed that the activity of genes which produce inducible enzymes was controlled by small proteins called **gene**

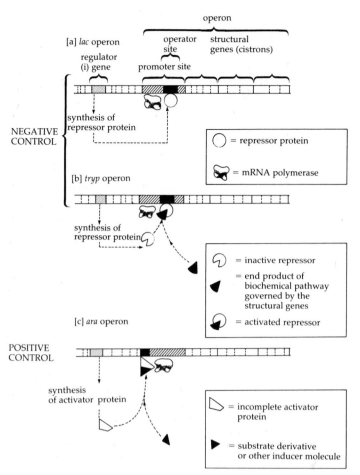

Fig. 3.6 *Operon systems in prokaryotes.* In **negative control** the regulatory (inhibitory) protein switches off the gene unless an inducer molecule, such as a substrate, is present (top diagram). A modification of this is shown in the middle diagram, where the end product of the reaction facilitates the action of the inhibitor. In **positive control** (bottom diagram), a regulatory protein (activator) switches the gene on. In order to do so, it may first have to complex with an appropriate substrate or hormone molecule. The last system is thought to be most similar to the system of control found in eukaryotes, though operons as such (orderly sequences of coordinately induced genes) do not exist in eukaryotes.

**regulators**. Some such genes are specifically inactivated by **repressors**, whereas others are switched on by **activators**. Repressors block mRNA synthesis by preventing the key enzyme *mRNA polymerase* from binding to DNA and transcribing it into mRNA. Activators promote synthesis by facilitating binding. Both work by attaching to an **operator site** next to the **structural genes** which code for the enzymes (Fig. 3.6).

In eukaryotes, specific gene repressors seem to be absent, although whole blocks of genes can be switched off by wrapping the DNA into a tightly coiled form (**heterochromatin**) which is inaccessible to gene-activating molecules and mRNA polymerase. Different blocks of genes tend to be switched off in histologically different cells. This may help to explain the stable differences which exist between, say, nerve and muscle cells.

**Gene activation** (positive control) is the preferred mechanism of transcriptional control in eukaryotes.

Since probably less than 5% of the DNA is active in any particular eukaryotic cell, natural selection has clearly favoured a system which specifically activates the few genes that are required rather than a system which *inactivates* the many that are not. The activation of eukaryotic genes by steroid hormones is illustrative. Steroid hormones are membrane soluble and pass into the cytosol of all cells. If a cell is in a tissue which responds to a particular hormone, then a **receptor protein** will be present. The two form a complex which then activates those genes that bring about the response characteristic of the hormone (Fig. 3.7). Any genes wrapped up into heterochromatin will not, of course, be activated, and those genes which are not wrapped up will *only* be activated if they contain an appropriate site to which the complex can bind.

---

**Q3** (i) What is a steroid hormone? Name one.
  (ii) How does the action of a steroid hormone in Fig. 3.7 differ from that of a protein hormone such as adrenalin?

---

It will be apparent that the components of transcriptional control can be put together in a variety of ways for the purposes of regulating mRNA synthesis.

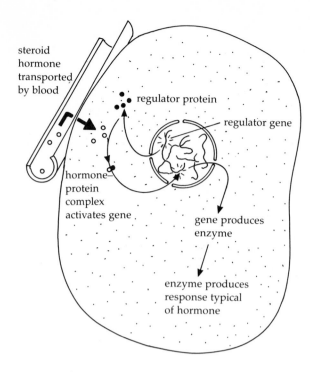

Fig. 3.7 *Gene induction by steroid hormones*

---

**Catabolite repression**

A catabolite is a substance used for respiration, e.g. glucose. It has been found that the production of the lactose-degrading enzymes in *E. coli* does not *just* depend upon whether an inducer (e.g. lactose) is present in the culture medium. In other words, Fig. 3.6(a) is an oversimplification. If glucose and lactose are present together, glucose inhibits the induction of the lactose-digesting enzymes. Hence the term *catabolite repression*. The mechanism appears to be as follows:

(i) In the absence of glucose, cAMP levels are high; in the presence of glucose, they are low.
(ii) cAMP at high concentrations binds to a protein called **catabolite gene activator protein** (CAP).
(iii) CAP then binds close to the operator site and assists *RNA polymerase* to bind to the DNA.

Hence transcription of lactose-degrading enzymes in *E. coli* depends on two things: the presence of lactose, and the presence of cAMP–CAP molecules (Fig. 3.8). Such a system makes considerable sense. Glucose is a more efficient energy source than lactose. So if both are present in a culture medium it would be an advantage to use the glucose first (before anything else does!). Catabolite repression by glucose is now known to be effective against a variety of sugars in addition to lactose. While in principle a similar system could operate in eukaryotes, there is little evidence that it does so.

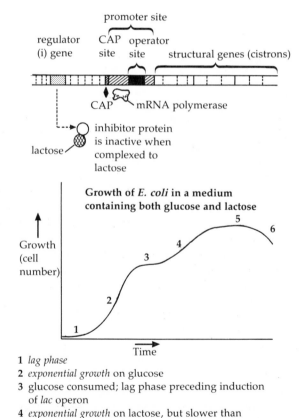

1 *lag phase*
2 *exponential growth* on glucose
3 glucose consumed; lag phase preceding induction of *lac* operon
4 *exponential growth* on lactose, but slower than on glucose
5 *stationary phase:* nutrients exhausted
6 *degenerate phase:* cells dying
The term **diauxic growth** is used to describe a situation where there are two distinct growth phases, as here (**2** and **4**).

Fig. 3.8 *CAP binding protein and catabolite repression*

*Translational control*

Two alternatives could in principle control the amount of mRNA translated into protein:
(i)   the machinery (mRNA, ribosomes) could remain 'dormant' unless specifically activated;
(ii)  the lifetime of mRNA in the cytoplasm could be regulated.
Translational control has only been confirmed in a few cases and the first alternative seems to be more important. For example, in fertilised eggs and germinating seeds, protein synthesis begins almost immediately, well before appreciable amounts of DNA are transcribed. The underlying control mechanism is obscure.

While the rate of mRNA degradation is certainly known to vary in different cells and tissues, the biological significance of this is uncertain.

### 3.3.2  Destruction

The rate at which an enzyme is destroyed will also affect the quantity present. The half-life of enzymes varies from a few minutes to several days, and is influenced by a number of environmental factors. The presence of substrate, for example, increases enzyme stability. To what extent enzyme destruction is controlled by the cell, if at all, is uncertain.

### 3.4  ENZYME LOCATION

Bacterial cells are so small (1–5 $\mu m^3$) that simple diffusion throughout the cell ensures that substrate and enzyme molecules collide sufficiently often for reactions to proceed at an appropriate rate. Eukaryotic cells are between about $10^3$ and $10^5$ times bigger. Being so much larger, simple diffusion throughout the whole cell would be inadequate. Consequently, eukaryotes have evolved an elaborate system of **membrane-bound organelles** such as endoplasmic reticulum, Golgi-bodies, lysosomes etc. (see *The Eukaryotic Cell*, Chapter 4, in this series). The organelles compartmentalise reactions, and in doing so increase metabolism in a number of ways:
(i)   By maintaining high substrate and enzyme concentrations inside a small organelle, the chances of collisions between the two are greater. Many components of respiration, for example, are confined to mitochondria. Those of photosynthesis are confined to chloroplasts.
(ii)  By the formation of **multi-enzyme complexes**. If metabolically related enzymes associate, the product of the first enzyme can easily be passed on to the second enzyme, and so on. A specific example is *pyruvate dehydrogenase*, found in mito-chondria, which prepares the breakdown products of glycolysis for aerobic respiration. Other examples occur in chloroplasts and endoplasmic reticula.
(iii) By the location of enzymes within membranes. Fat-soluble substrates will dissolve in the membrane. This improves the chances of a collision with membrane-bound enzymes because random movement of the molecules is restricted from three dimensions to two.
(iv)  By separating substrate and product. Chemi-osmosis (Chapter 4), which drives the synthesis ATP in mitochondria and chloroplasts, requires a physical separation of reactants. Phospholipid membranes provide the necessary barrier.

Compartmentalisation of reactions also assists in regulation:
(i)   Dangerous enzymes can be isolated from the rest of the cell. The hydrolytic enzymes in lysosomes, for example, would rapidly destroy cells if they were free in the cytosol.
(ii)  Dangerous materials can be isolated from the rest of the cell. For example, toxic hydrogen peroxide is made and destroyed (by *catalase*) within the confines of peroxisomes.

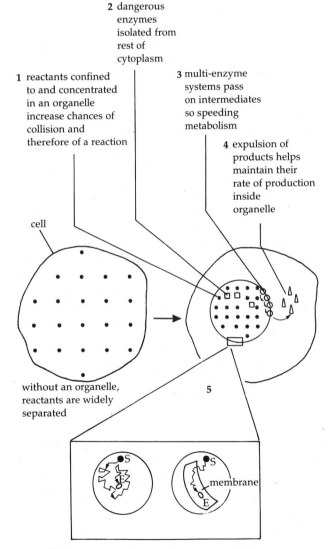

Inset: Without a membrane, the chances of a collision between substrate (S) and enzyme (E) are less, because S diffuses in three dimensions. However, if S is lipid soluble, the 'time-to-target' is reduced by $10\times - 100\times$ because S is only diffusing in two dimensions after hitting the membrane.

Fig. 3.9 *The contribution of membranes to metabolic control*

## 3.5 ENZYME TYPE

### Isozymes

It is recalled (Chapter 2.1.1) that enzymes are named according to their function, *not* by their chemical composition. It has been found that enzymes with a particular function can exist in slightly different molecular forms. Thus whereas any starch-digesting enzyme is always called α-*amylase*, the α-*amylase* in a germinating pea may, in terms of molecular structure, be slightly different from that in the human gut. Even within a single organism or cell, functionally equivalent enzymes with slightly different chemical properties may exist. The latter are called **isozymes** (syn.: **isoenzymes**). Isozymes make possible subtle modifications to metabolism (Q4).

---

**Q4** In Fig. 3.10 the conversion of A to B is regulated by two enzymes, $e_1$ and $e_2$. B is subsequently converted to end-products C and D via other intermediate reactions.

(i) How can the formation of C be maintained if D builds up to an unacceptably high level?

(ii) If the concentration of C increases excessively:
 (a) how can its own production be reduced?
 (b) how can the production of D be maintained?

*Example*  amino acid synthesis
A aspartate
B aspartyl phosphate
$e_1$, $e_2$ isozymes of aspartate kinase
C lysine
D threonine
Location: *E. coli*

---

### Irreversible reactions

An enzyme always catalyses a reaction both ways, but some reactions are so exergonic that *at equilibrium* a trivial amount of substrate may be left. Such reactions are described as irreversible. Thus:

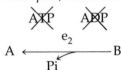

$$A \xrightarrow{\quad\quad} B \ (\Delta G' \text{ very negative})$$

---

*Example*
A glucose
B glucose-6-phosphate
$e_1$ *hexokinase*
Location: most tissues

---

Irreversible reactions are often coupled to ATP utilisation, as in the above example. Indeed it is this coupling which makes them irreversible. If the ATP is *uncoupled*, the reaction can in effect, go backwards:

$$A \xleftarrow{\quad e_2 \quad} B$$

---

*Example*
A glucose
B glucose-6-phosphate
$e_2$ *glucose phosphatase*
Location: liver, kidney

---

It is important to note, however, that by excluding ATP/ADP the 'backwards' reaction is *not exactly* a reverse of the forward reaction. As a result, a different enzyme is used: one which has no binding site for ATP/ADP. A variation on this pattern is given below (Q5).

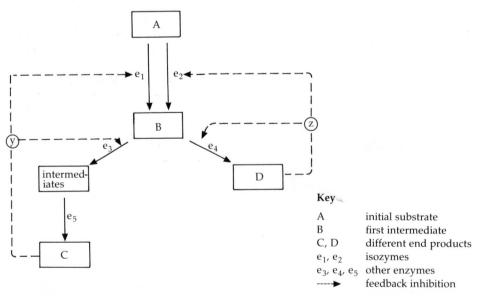

Fig. 3.10 *Isozymes*. **Isozymes** (syn.: **isoenzymes**) are structurally and functionally related enzymes existing in subtly different forms and with slightly different properties.

**Q5** Study the diagram below. The breakdown of PEP to pyruvate is extremely exergonic: so much so that it is coupled to ATP *synthesis*. (It is one of only two points where ATP is made in the first stage of respiration (glycolysis).) Nevertheless, glucose *can* be formed from organic acids, such as pyruvate, as indicated in the diagram. This happens in liver after exercise. How is this re-synthesis *energetically* possible?

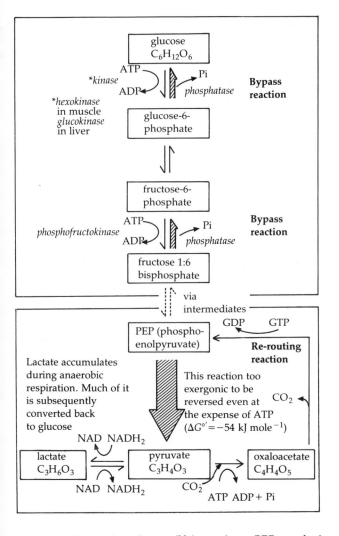

Fig. 3.11 *Overcoming 'irreversible' reactions: PEP synthesis during gluconeogenesis (glucose synthesis from lactate).* The text preceding Q5 explains the upper part of the diagram (bypass reactions). Q5 applies specifically to the lower part of the diagram (re-routing reaction).

The ability to employ different enzymes for by-passing or re-routing 'irreversible' reactions is one of the most powerful weapons the cell possesses for metabolic regulation. We shall meet this strategy several times in the following chapters.

---

**Study guide**

*Vocabulary*

Explain what is meant by:
covalent modification     product inhibition
isozyme                          pacemaker
negative feedback         secondary messenger

*Review Questions*

1 The phosphorylation of the amino acid aspartate at the expense of ATP is an important first step in a pathway leading to the synthesis of various amino acids including lysine and threonine. Enzymes capable of this phosphorylation are called *aspartate kinases*. A solution showing *aspartate kinase* activity was isolated from a culture of the bacterium *E. coli* K12.
The effects of (a) lysine alone, (b) threonine alone, and (c) threonine and lysine together on the activity of the enzyme were investigated.

| Concentration of additive (mM) | | 0 | 1 | 2 | 4 | 12 |
|---|---|---|---|---|---|---|
| Percentage activity of enzyme | with lysine | 100 | 40 | 50 | 50 | 55 |
| | with threonine | 100 | 30 | 35 | 55 | 55 |
| | with lysine and threonine | 100 | 70 | 90 | 95 | 95 |

(i) Outline how in principle this experiment might be performed and what precautions in the experimental procedure might be necessary.
(ii) Graph the results, using the same pair of axes for the data.
(iii) What conclusions might be drawn from these results? What further experiments might be performed to test your explanation?

2 Under what circumstances would it be most appropriate to use control of (i) enzyme activity, (ii) transcription and (iii) translation as a means of metabolic regulation? Give reasons for your answer.

# Respiration

### SUMMARY

Respiration of sugars by glycolysis to lactate (or ethanol and $CO_2$) is non-oxidative and releases only a small amount of the energy potentially available. Oxidation, if it occurs, yields considerably more energy. In both cases a proportion of the energy released is captured in the useful form of ATP. In oxidative respiration, ATP synthesis occurs by means of an electron-powered proton pump which drives the reaction:

$$H^+ + ADP^{3-} + Pi^{2-} \rightarrow ATP^{4-} + H_2O$$

Respiration is mostly controlled by allosteric regulators through negative feedback systems.

## 4.1 GLYCOLYSIS: NON-OXIDATIVE RESPIRATION

The respiration of carbohydrate, particularly glucose and its polymers, starch and glycogen, provides a substantial part of the energy required by living organisms. The first step in the respiration of glucose (and, in the absence of suitable oxidising agents, the only step) is called **glycolysis**, which literally means 'splitting glucose'. It is an appropriate name, since glucose ($C_6H_{12}O_6$) is converted to lactic acid ($C_3H_6O_3$) or, in the case of plants, to ethanol ($C_2H_5OH$) and $CO_2$.

The botanical variant is usually specifically called **alcoholic fermentation**. Since it is identical in all major respects with glycolysis except at the very end of the pathway, the following account may be taken as applicable to both except where specified. There is no nett oxidation of glucose during glycolysis, and relatively little energy is made available for the synthesis of ATP.

Table 4.1 enables some important conclusions to be drawn about glycolysis and alcoholic fermentation, even without any knowledge of the intermediate steps (Q1–Q6).

**Q1** If $\Delta G^{o'} = -30$ kJ mole$^{-1}$ for ATP hydrolysis (ATP $\rightarrow$ ADP + Pi), what is the value of $\Delta G^{o'}$ for ATP synthesis? Explain.

From Table 4.1:

**Q2** How many moles of ATP are generated per mole of glucose?

**Q3** How many kilojoules of energy are therefore captured as ATP during glycolysis?

**Q4** What percentage of the total energy liberated during (i) glycolysis and (ii) alcoholic fermentation is therefore captured in the form of ATP?

**Q5** What happens to the energy which is *not* captured as ATP? What percentage of the total is this?

**Q6** Why is caution needed when interpreting calculations based on $\Delta G^{o'}$?

Although glycolysis as a whole is exergonic, not every step in the pathway provides energy for ATP synthesis. Indeed the pathway is a mixture of some endergonic and some highly exergonic reactions. It falls into five main phases:

### 1 Glucose activation

Two intrinsically very endergonic reactions ((i) and (iii)) which occur at the beginning of glycolysis (Fig. 4.1, dotted line) are made favourable (exergonic) by coupling them to ATP hydrolysis.

Table 4.1 *Glycolysis and alcoholic fermentation*

| | |
|---|---|
| *Location:* | Cytosol |
| *Steps:* | 10 (glycolysis) or 11 (alcoholic fermentation), all enzymically mediated |
| *Substrate:* | Glucose (or other hexose sugars); $O_2$ is *not* consumed |
| *Products:* | Lactic acid (glycolysis); ethanol and $CO_2$ (alcoholic fermentation). $H_2O$ is *not* produced |
| *ATPs generated:* | Two per glucose consumed. Formed by substrate-linked phosphorylation |
| *Summary:* | *Glycolysis:* $C_6H_{12}O_6 \rightarrow 2C_3H_6O_3$ ($\Delta G^{o} = -200$ kJ mole$^{-1}$) |
| | *Alcoholic fermentation:* $C_6H_{12}O_6 \rightarrow 2C_2H_5OH + 2CO_2$ ($\Delta G^{o} = -210$ kJ mole$^{-1}$) |

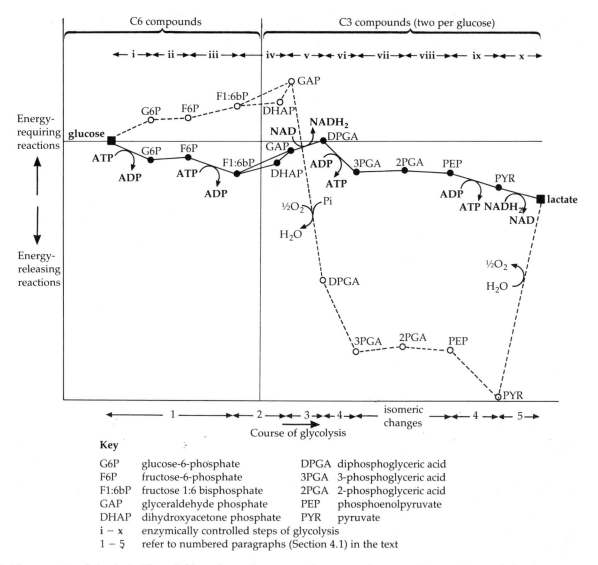

**Key**

| | | | |
|---|---|---|---|
| G6P | glucose-6-phosphate | DPGA | diphosphoglyceric acid |
| F6P | fructose-6-phosphate | 3PGA | 3-phosphoglyceric acid |
| F1:6bP | fructose 1:6 bisphosphate | 2PGA | 2-phosphoglyceric acid |
| GAP | glyceraldehyde phosphate | PEP | phosphoenolpyruvate |
| DHAP | dihydroxyacetone phosphate | PYR | pyruvate |
| i − x | enzymically controlled steps of glycolysis | | |
| 1 − 5 | refer to numbered paragraphs (Section 4.1) in the text | | |

Fig. 4.1 *The energetics of glycolysis.* The solid line shows the energy changes as they actually occur in coupled reactions involving ATP and NAD. The dotted line shows the energy changes which would occur without coupling.

---

**Q7** Name the substrates and products of these reactions.

**Q8** How many ATPs per glucose are needed to start the pathway?

**Q9** The enzymes concerned in reactions (i) and (iii) are called *hexokinase* and *phosphofructokinase*. What does the ending '-kinase' indicate?

---

ATP is said to be 'energising' the glucose molecule in these reactions or acting as a 'pump primer'. This may seem a peculiar idea, since glucose is itself supposed to be a source of chemical energy. Note, however, that the overall shape of the dotted graph (Fig. 4.1) is very similar to the activation barrier graph for a single reaction described previously (Fig. 2.2). The analogy is a fair one: the two ATPs used at the beginning of glycolysis are in a sense providing glucose with energy so that the 'activation energy barrier' for the pathway as a whole is reduced (Fig. 4.1, solid line).

**2 Splitting the sugar: C6 → 2 × C3**
When fructose bisphosphate ('activated glucose') is hydrolysed, it yields two slightly different phosphory-

lated molecules, an aldehyde (—CHO) and a ketone (≥CO). Only the aldehyde is used up in the next step; the ketone spontaneously forms aldehyde as the latter is consumed.

---

**Q10** Hydrolysis of the C6 sugar has a positive $\Delta G^{\circ\prime}$. How can the reaction possibly proceed?

---

**3 Oxidation**
Oxidation of the C3 aldehyde (—CHO) to an acid (—COOH) is, like all oxidations, strongly exergonic. As in most biological redox reactions, oxygen is not used directly, but instead hydrogen is removed by the electron (or hydrogen) carrier NAD to form $NADH_2$. This trapped hydrogen is a form of energy, sometimes called **reducing power** (Chapter 1.3.2).

In addition, some of the energy made available during the oxidation is used to build a second phosphate into the acid. Hence the nett result of the combined oxidation and phosphorylation is a key compound, diphosphoglyceric acid (**DPGA**).

## 4 Substrate-linked phosphorylation

If any substance in the cell merits the title 'high energy compound' it is DPGA, not ATP. In strongly exergonic reactions, DPGA is dephosphorylated, and ATP is produced:

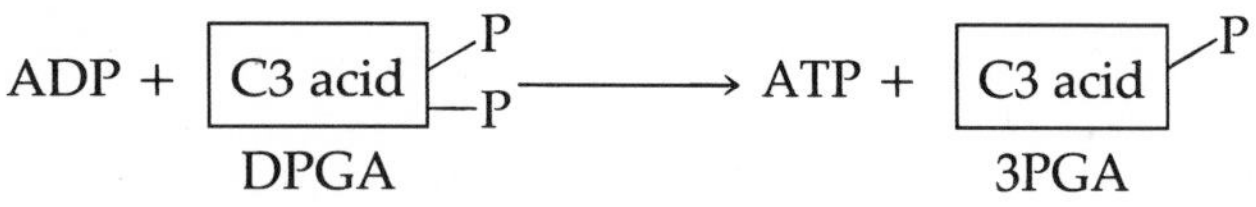

$$ADP + \boxed{\text{C3 acid}}_{P}^{P} \longrightarrow ATP + \boxed{\text{C3 acid}}^{P}$$
$$\qquad\quad \text{DPGA} \qquad\qquad\qquad\qquad \text{3PGA}$$

Phosphoglyceric acid (PGA) is then converted to a related acid, phosphoenol pyruvic acid (PEP):

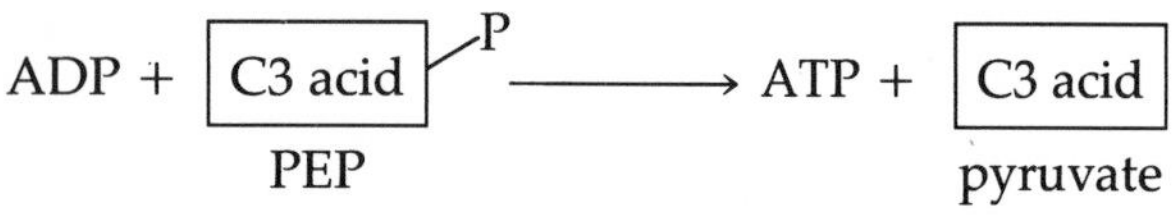

$$ADP + \boxed{\text{C3 acid}}^{P} \longrightarrow ATP + \boxed{\text{C3 acid}}$$
$$\qquad\quad \text{PEP} \qquad\qquad\qquad\qquad \text{pyruvate}$$

---

**Q11** Step 4 generates two ATPs per DPGA consumed. Step 1 consumes two ATPs per glucose. Yet Table 4.1 indicates a nett production of two ATPs per glucose. Explain this apparent contradiction.

---

The term **substrate-linked phosphorylation** means that ATP is generated by coupled reactions, as above. It is the only method of ATP synthesis in glycolysis and fermentation.

## 5 Regeneration of NAD

There is only a minute amount of NAD in the cell. Since step 3 converts this to $NADH_2$, glycolysis will come to an abrupt and fatal conclusion unless the '$H_2$' component is unloaded, and NAD is regenerated. In higher animals and plants there are two main possibilities.

(a) *Oxidation by pyruvate*
In the absence of oxygen, glycolysis goes to completion and the '$H_2$' of $NADH_2$ is unloaded onto pyruvate. In animals, lactic acid is produced. In the long term, lactic acid is toxic and must ultimately be removed (Chapter 6). In the short term, the reaction satisfies the greater priority of regenerating NAD:

$$NADH_2 + \underset{\text{(pyruvate)}}{C_3H_4O_3} \longrightarrow \underset{\text{(lactate)}}{C_3H_6O_3} + NAD$$

Lactic acid is similarly produced in some bacteria and plant tissues. In most plant tissues, however, the equivalent reaction is slightly different, producing an alcohol ($-CH_2OH$) via an aldehyde intermediate ($-CHO$): hence the term alcoholic fermentation. It is rather more wasteful than glycolysis, since an atom of organic carbon is lost as $CO_2$. (No doubt many people would regard this as a small price to pay for the resulting product!)

$$NADH_2 + \underset{\text{(pyruvate)}}{C_3H_4O_3} \xrightarrow[\text{(CH}_3\text{CHO)}]{\overset{CO_2}{\nearrow}\text{ via ethanal}} NAD + \underset{\text{(ethanol)}}{C_2H_5OH}$$

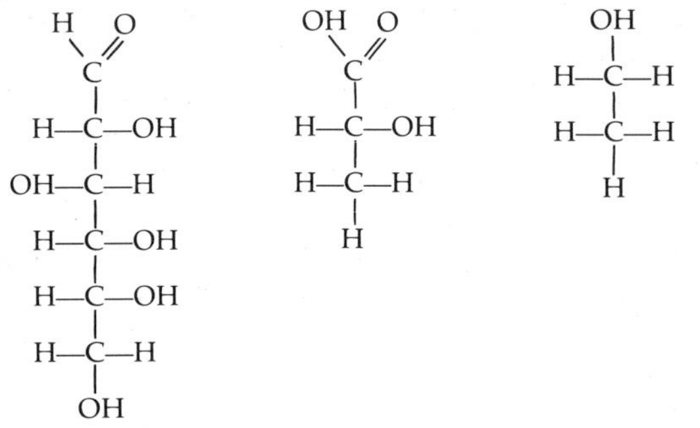

Fig. 4.2 *Glucose, lactate and ethanol*

(b) *Oxidation by molecular oxygen*
Suitable oxidising agents from the environment may also be used to accept the '$H_2$' and thus regenerate NAD. The most important of these (and in higher organisms the only possibility) is molecular oxygen itself. Moreover, when oxygen is available, glycolysis is interrupted at pyruvate. The latter then passes into the mitochondria where it is broken down via the tricarboxylic acid cycle to $CO_2$ and $H_2O$ (Section 4.3).

### 4.1.1 Control

The glycolytic enzymes are constitutive (Chapter 3.3.1) and the pathway is controlled by regulating enzyme activity. Significant differences exist between the control systems of different cells: that in muscle is comparatively straightforward and can be taken as illustrative.

*Muscle*

In muscle, *phosphofructokinase* (PFK) is the main pace-maker (Chapter 3.2.1). It acts like a valve: close it down and the supply of fuel through the system slows; open it up and the rate of energy production increases. The activity of PFK is determined by a number of factors:

(i) *The energy status of the cell.* This is particularly important for minute-to-minute regulation. Since the primary effect of glycolysis is the conversion of ADP to ATP, we might expect that these substances would affect PFK. It is not quite so simple. For a start, any tendency for the level of ATP to drop is counteracted by the action of an enzyme called *adenylate kinase*:

$$2ADP \xrightarrow[\text{kinase}]{\text{adenylate}} AMP + ATP$$

Hence a demand for energy is marked not so much by a drop in ATP concentration, but by a rise in AMP concentration. The latter is a potent activator of PFK, and so causes glycolysis to speed up. Conversely, ATP is an inhibitor of PFK, so any tendency for the ATP concentration to rise causes the rate of glycolysis to slow down.

(ii) *Glucose level.* A rising concentration of glucose, however caused, increases the level of F6P (Fig. 4.3). This is converted into *two* forms of fructose bisphosphate:

(a) Fructose 1:6 bisphosphate: the 'normal' metabolite of glycolysis (Figs. 3.11, 4.1).

(b) Fructose 2:6 bisphosphate: this is only found in small amounts, but is a very potent activator of PFK. Hence, a rise in glucose leads to a rise in F2:6bP, which in turn activates PFK. The nett result is that glycolysis speeds up so helping to reduce the cellular level of glucose to within its normal limits. Conversely, a drop in cellular glucose results in a decrease in F2:6bP, and glycolysis slows.

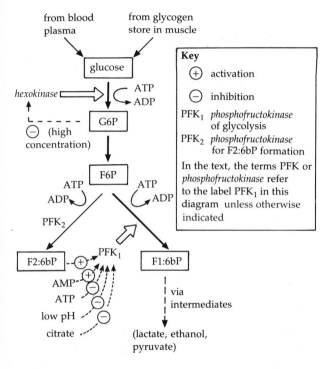

Fig. 4.3 *The control of glycolysis*

(iii) *Fatty acid breakdown.* The consumption of fats rather than carbohydrates for energy leads to a rise in fatty acid level and a rise in citrate. The latter is an inhibitor of PFK. In other words, oxidation of fatty acids switches off (is antagonistic to) oxidation of glucose.

(iv) *pH.* High lactate levels, which may occur during exercise, drive down the pH of the cytoplasm (acidosis), leading to inhibition of PFK.

*Hexokinase* is regulated in tandem with *PFK*: G6P accumulates as *PFK* is inhibited, leading to product inhibition of *hexokinase*. Glucose does not simply accumulate in the cell, however, because its rate of absorption will decrease, and any excess will be converted into glycogen.

The variety of factors which influence the rate of glycolysis in muscle reflects the variety of internal conditions which must be maintained within close limits. It also reflects the variety of circumstances under which muscle may have to operate. It is therefore not surprising that levels of ATP, AMP, glucose (via

hormones, starvation or exercise), citrate and $H^+$ all contribute to regulation. Anything simpler would be inadequate.

---

**The pentose phosphate pathway**

Glucose may be broken down by pathways other than glycolysis. In most plants, and in some animal tissues such as mammary glands, substantial quantities of glucose are degraded by the **pentose phosphate pathway**.

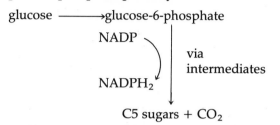

This pathway generates no ATP, and results in a loss of organic carbon. However, it has two valuable functions:

(i) It conserves energy as reducing power in the form of $NADPH_2$. This is used to synthesise fatty acids, cholesterol, carotene and other lipids (Chapter 6).

(ii) It generates C5 sugars such as ribose for the synthesis of nucleotides and polynucleotides like ATP, NAD, RNA and DNA.

---

### 4.2 OXIDATIVE RESPIRATION

**Q12** Define the term *redox reaction*.

**Q13** During an oxidation the substance which accepts (gains) electrons is always: (i) oxygen; (ii) hydrogen; (iii) reduced; (iv) oxidised.

As stressed previously, when electrons pass *down* Table 1.1 from a reducing agent to an oxidising agent, the reaction is exergonic, often markedly so. Oxidation therefore provides further potential for ATP synthesis. The commonest case is where oxygen itself is the electron acceptor, in which case respiration is described as **aerobic respiration** (Fig. 4.4).

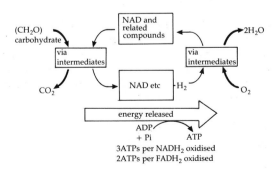

Fig. 4.4 *Aerobic oxidative respiration.* Oxidation of $NADH_2$/$FADH_2$ is coupled to ATP synthesis, the former being 'worth' 3ATPs, and the latter 2ATPs, as shown.

There are several substances which can substitute for oxygen in particular circumstances. Examples include $NO_3^-$, $SO_4^{2-}$ and fumarate. In such cases respiration is still oxidative, but since molecular oxygen is not used it is **anaerobic**.

Table 4.2 *Types of oxidative respiration.* The examples shown are illustrative, and a number of other patterns exist. (DPGA, diphosphoglyceric acid.)

| Oxidative reaction | Organism |
|---|---|
| **Aerobic** | |
| Glucose $\longrightarrow CO_2$ $O_2 \rightarrow H_2O$ | Most organisms |
| **Anaerobic** | |
| Glucose $\longrightarrow CO_2$ $NO_3^- \rightarrow N_2 + H_2O$ | Denitrifying bacteria e.g. *Paracoccus* |
| Glucose $\longrightarrow CO_2$ $SO_4^{2-} \rightarrow H_2S + H_2O$ | Putrifying bacteria e.g. *Desulphovibrio* |
| Glucose $\longrightarrow$ DPGA $C_4H_4O_4 \rightarrow C_4H_6O_4$ Fumarate    Succinate | Parasitic worms e.g. *Monieza expansa* (a tapeworm) |

**Q14** In Chapter 1 (Q5) you calculated $\Delta G^{o'}$ for the oxidation of $NADH_2$ by molecular oxygen. Using Table 1.1, calculate $\Delta G^{o'}$ for the oxidation of $NADH_2$ by $NO_3^-$. (Use the same equation, $\Delta G^{o'} = -zFE$.)
Is as much energy available for ATP synthesis when $NO_3^-$ replaces $O_2$? In your calculations, what determines the magnitude of the energy available?

**Q15** Which substance, $O_2$ or $NO_3^-$, should an organism use if it needs to generate as much ATP as possible?
Some soil bacteria like *Pseudomonas* can switch from using oxygen to $NO_3^-$ if the former is not available. They are called **facultative anaerobes**. Anaerobic soil conditions are promoted by waterlogging, which drives out the air. In such conditions, *Pseudomonas* reduces soil fertility by acting as a **denitrifying bacterium**, since the oxygen substitute $NO_3^-$ is reduced to $NO_2^-$ or $N_2(g)$.

Since oxidation of $NADH_2$ by these 'oxygen substitutes' generates less ATP, oxygen substitution is principally confined to micro-organisms and parasitic worms exploiting restricted anaerobic habitats.

### 4.3 AEROBIC RESPIRATION IN EUKARYOTES

In eukaryotes, aerobic respiration takes place exclusively in **mitochondria** (Fig. 4.5). In this process, simple

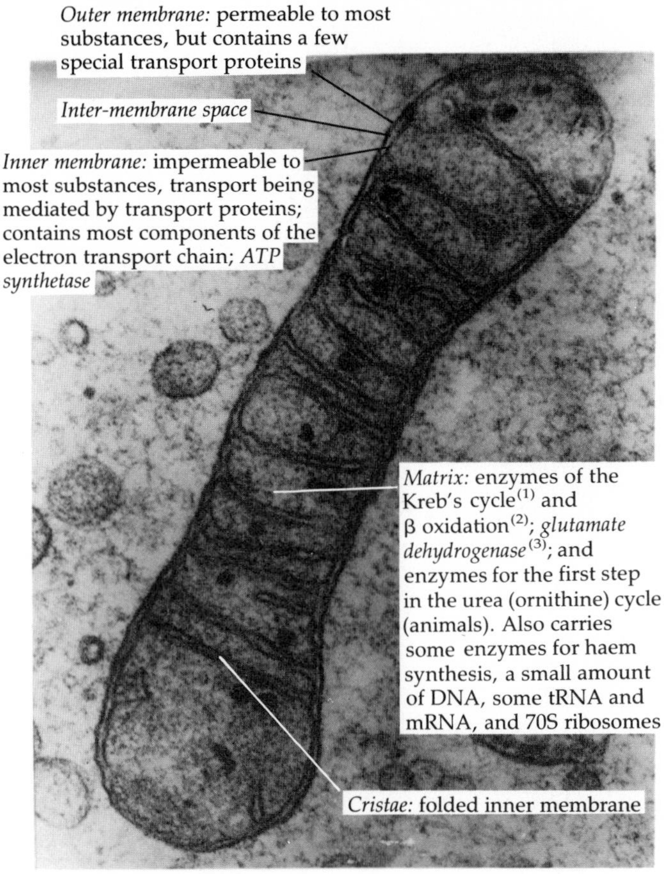

*Outer membrane:* permeable to most substances, but contains a few special transport proteins

*Inter-membrane space*

*Inner membrane:* impermeable to most substances, transport being mediated by transport proteins; contains most components of the electron transport chain; *ATP synthetase*

*Matrix:* enzymes of the Kreb's cycle[1] and β oxidation[2]; *glutamate dehydrogenase*[3]; and enzymes for the first step in the urea (ornithine) cycle (animals). Also carries some enzymes for haem synthesis, a small amount of DNA, some tRNA and mRNA, and 70S ribosomes

*Cristae:* folded inner membrane

| | |
|---|---|
| Location | Cytoplasm of all aerobically respiring cells |
| Number | Up to 1000 per cell |
| Shape | Extremely variable: ovoid to snakelike filaments |
| Size | From about 1 μm diameter to filaments 1 μm × 40 μm |
| Major functions | (i) ATP generation by oxidative phosphorylation |
| | (ii) Heat production |
| | (iii) Miscellaneous syntheses |
| Other features | Self replicatory |

Fig. 4.5 *Structure and function in mitochondria.* [1]Used in the oxidation and decarboxylation of organic acids derived from carbohydrates (by glycolysis), fats and proteins. [2]Used for the breakdown of fatty acids. [3]Used for the deamination of amino acids.

organic (**carboxylic**) acids derived from carbohydrates, fats and proteins are metabolised to $CO_2$ and $H_2O$, making available substantial quantities of energy for ATP synthesis. The fundamental mechanism is a series of eight enzymically controlled steps known as the **tricarboxylic acid cycle** (Fig. 4.6) (syn.: Krebs cycle, citric acid cycle). $CO_2$ is released by decarboxylation reactions; $H_2O$ is formed as a result of oxidation reactions. The hydrogen component in the $H_2O$ originates from the organic acids, whereas the oxygen originates from the atmosphere. It is during the transport of hydrogen to oxygen by an **electron transport chain** that most of the energy for ATP synthesis becomes available. The process is therefore commonly called **oxidative phosphorylation**. For the most part the enzymes of the Krebs cycle are found in the gelati-

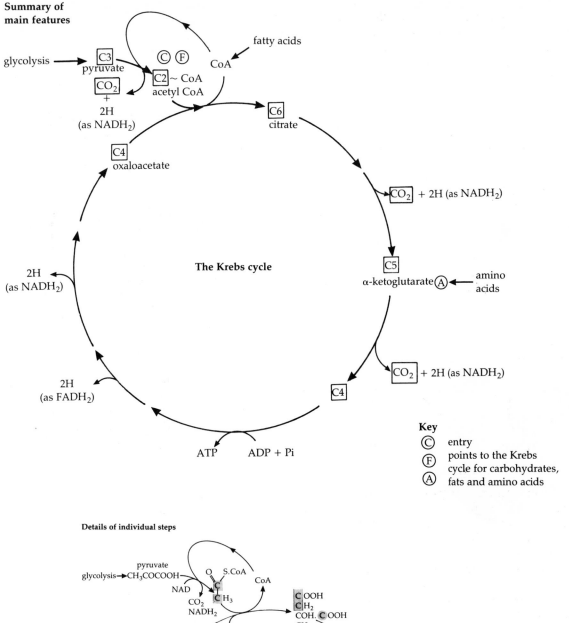

**Summary of main features**

The Krebs cycle

**Key**

Ⓒ entry
Ⓕ points to the Krebs cycle for carbohydrates,
Ⓐ fats and amino acids

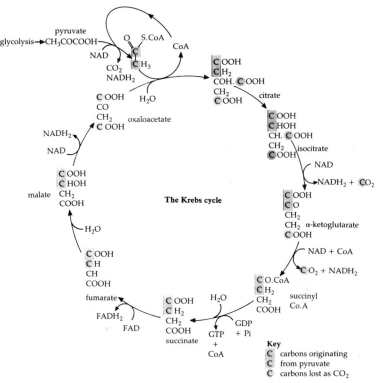

Details of individual steps

The Krebs cycle

**Key**

**C** carbons originating from pyruvate
**C** carbons lost as $CO_2$

Fig. 4.6 *Aerobic respiration in mitochondria: Part 1 – the Krebs Cycle.* The significant points are embodied in the upper diagram. The lower diagram is included merely for reference: there is little to be gained from memorising the detail shown.

nous matrix of the mitochondrion, whilst the components of the electron transport chain and **ATPase** (*ATP synthetase*) are confined to the **mitochondrial inner membrane**.

### 4.3.1 Entry points

The principal points of entry into the Krebs cycle are **acetyl CoA** and **α-ketoglutarate** (Fig. 4.7). Acetyl CoA is by far the more important entry point. It can be formed in two ways:

(i) *via glycolysis*
Glycolysis generates two molecules of pyruvate per glucose. If fat is being respired one molecule of pyruvate will also be generated for each lipid molecule consumed (this will originate from the glycerol sub-unit in the lipid). Whatever the source, pyruvate now diffuses from the cytosol into the mitochondrion. By a series of reactions involving a multi-enzyme complex called *pyruvate dehydrogenase*, pyruvate combines with

coenzyme A to form acetyl CoA at the expense of one atom of organic carbon:

$$\boxed{C3} + CoA \xrightarrow[\text{pyruvate dehydrogenase}]{\text{NAD} \quad \text{NADH}_2} \boxed{C2} -CoA + CO_2$$

pyruvate — acetyl COA: high energy 'carbon carrier'

A small amount of pyruvate may also be formed directly in the mitochondria by deamination of the amino acid alanine. This is similarly converted to acetyl CoA.

(ii) *via β-oxidation*
In this pathway fatty acids are, in effect, chopped into chunks of acetyl CoA. Details of β-oxidation are given in Chapter 6.

An input of α-ketoglutarate to the cycle is quantitatively important when proteins are being respired. It is produced by **deamination** of glutamate which is in

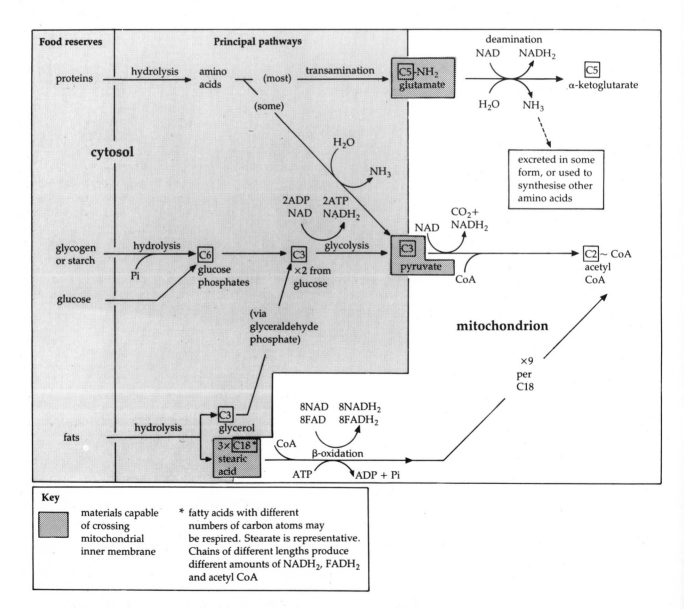

Fig. 4.7 *Entry points to the Krebs cycle*

turn formed (by **transamination**) from a variety of amino acids. The mechanism is summarised in Fig. 4.7 (see also Chapter 6.3).

### 4.3.2 Decarboxylation

Acetyl CoA (C2) reacts with **oxaloacetate** (C4) to produce **citrate** (C6) and release CoA. The latter is recycled to pick up more pyruvate or fatty acid fragments. Citrate is degraded back to oxaloacetate via several intermediate reactions, so enabling the Krebs cycle to continue (Fig. 4.6).

---

**Q16** How many carbon atoms must be released from citrate in order to regenerate oxaloacetate, i.e. how many decarboxylations ($CO_2$-releasing steps) occur?

**Q17** How many turns of the Krebs cycle are therefore needed to account for the number of carbons in the acetyl fragment of acetyl CoA?

**Q18** Why are *2 acetyl CoA (C2) ≡ 2 pyruvate (C3) ≡ 1 glucose (C6)*?

**Q19** How many turns of the Krebs cycle are therefore needed to account for the number of carbon atoms in glucose?

---

As shown in the detailed inset of Fig. 4.6, an acetyl group entering the Krebs cycle does not leave as $CO_2$ in the same 'turn', but in the next one.

### 4.3.3 Phosphorylation, oxidation and control

Each turn of the Krebs cycle generates one ATP by substrate-linked phosphorylation, but this is an insignificant amount compared with the quantity generated by oxidative phosphorylation. The latter involves two steps:

(i) Starting, for convenience, from pyruvate, five oxidations occur during the Krebs cycle, which build up a supply of $NADH_2$ and $FADH_2$:

$$\underset{\text{pyruvate}}{C_3H_4O_3} + 3H_2O \xrightarrow[\text{and Krebs}]{\text{via acetyl CoA}} 3CO_2 + 10H$$

$$4NAD \overset{\frown}{\phantom{xx}} 4NADH_2$$
$$FAD \phantom{xxx} FADH_2$$

---

**Q20** Identify the points in Fig. 4.6 where these oxidations occur.

---

(ii) The reduced dinucleotides are themselves now oxidised in strongly exergonic reactions which are coupled to ATP synthesis: hence the term 'oxidative phosphorylation'.

$$NADH_2 + \tfrac{1}{2}O_2 \longrightarrow NAD + H_2O$$
$$(\Delta G^{o'} = -220 \text{ kJ mole}^{-1})$$
$$3ADP + 3Pi \overset{\frown}{\longrightarrow} 3ATP$$

$$FADH_2 + \tfrac{1}{2}O_2 \longrightarrow FAD + H_2O$$
$$(\Delta G^{o'} = -152 \text{ kJ mole}^{-1})$$
$$2ADP + 2Pi \overset{\frown}{\longrightarrow} 2ATP$$

A supply of NAD (and FAD) is clearly essential for the Krebs cycle to operate. This is normally maintained by oxidation of its reduced form by molecular oxygen. If the latter is in short supply, the cycle will slow down, and pyruvate may be pushed towards lactate or alcohol synthesis. There are other control points which allow 'fine tuning'. For example, acetyl CoA inhibits its own production at high concentration.

*The efficiency of aerobic respiration*

In Section 4.1.1, calculations were made concerning the efficiency of glycolysis. Similar calculations can be made about aerobic respiration (RQ1).

## 4.4 CHEMIOSMOSIS IN MITOCHONDRIA

Oxidative phosphorylation involves the participation of a system of intermediates, the electron transport chain, consisting mostly of haem-containing proteins called **cytochromes**. These substances are embedded in the inner mitochondrial membrane (Fig. 4.8). The mechanism by which oxidative phosphorylation occurs is called **chemiosmosis** (osmos: to push (protons)).

Mitchell's chemiosmotic theory (1961) explains both oxidative respiration (mitochondria) and photosynthetic phosphorylation (chloroplasts, Chapter 5). In essence it proposes that ATP synthesis is simply a reversal of hydrolysis in which a locally high concentration of protons ($H^+$) forces an otherwise unfavourable reaction towards synthesis:

$$ATP^{4-} + H_2O \underset{\phantom{hydrolysis}}{\overset{\text{hydrolysis}}{\rightleftharpoons}} ADP^{3-} + Pi^{2-} + H^+$$
$$(\Delta G^{o'} = -30 \text{ kJ mole}^{-1})$$

Chemiosmosis as applied to mitochondria involves four independent postulates:
(i) *The mitochondrial inner membrane is generally impermeable to $H^+$*
This has been confirmed (Q21, Q22).

---

**Q21** If the electron transport chain is active, the pH of the mitochondrial matrix is higher (more alkaline) than that of the intermembrane space or cytosol. Explain this in terms of $H^+$ concentration.

**Q22** Drugs which punch holes in membranes cause this pH difference to disappear. How is this result consistent with the first postulate?

---

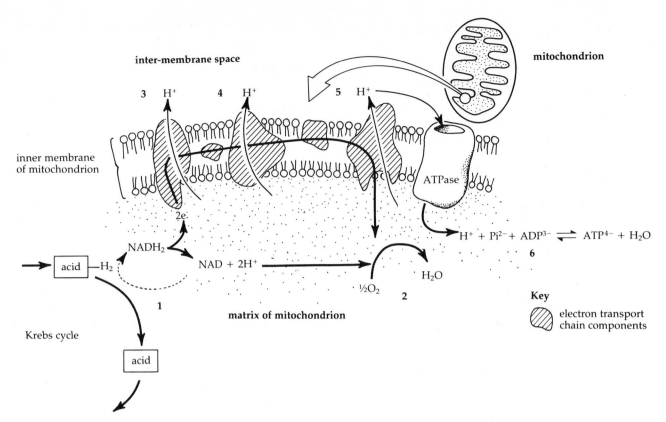

Fig. 4.8 *Aerobic respiration: Part II — oxidative phosphorylation by chemiosmosis*

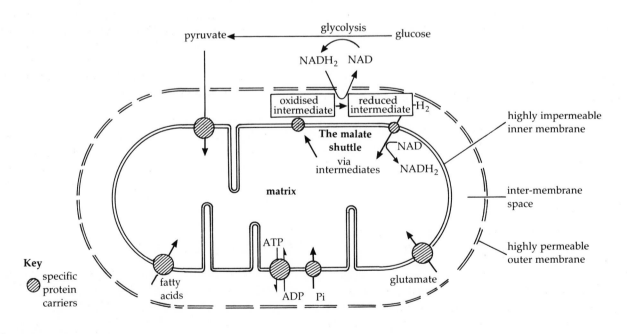

**The mitochondrial inner membrane**
Evidence suggests that the inner membrane is not only impermeable to $H^+$, but to a wide range of other substances as well. These include ATP, ADP, $NADH_2$ and pyruvate (Fig. 4.9).

Fig. 4.9 *Shuttle systems through the inner mitochondrial membrane.* Carrier systems are frequently involved in moving materials across the inner membrane. Entry of one substance, such as ADP, is often linked to the export of another, such as ATP. $NAD/NADH_2$ are themselves quite impermeable but 'reducing equivalents' may be transported across using a shuttle system. Since the reduced carrier is usually malate, the system is often called the **malate shuttle**.

(ii) *The electron transport chain pumps $H^+$ ions out of the matrix into the intermembrane space*

As an electron passes from $NADH_2$ along the chain towards $O_2$, some intermediate steps require protons whilst others expel them. The second postulate therefore demands a careful arrangement of proton-requiring and proton-expelling intermediates, so that a nett outward pumping can occur. Biochemical and electron micrographic evidence confirms both the pumping action and the predicted asymmetry of the electron transport chain.

(iii) *The accumulation of $H^+$ on the cytosol side of the inner membrane is a source of potential energy for ATP synthesis*

As $H^+$ is pumped out, a **membrane potential**, $\Delta\psi$ (voltage difference across the membrane), is produced due to a **charge separation** which comes from an excess of positive ($H^+$) ions on the outside.

In addition, whilst the matrix = pH 8, the intermembrane space = pH 7, implying a $10\times$ greater concentration of $H^+$ ions on the outside. This **pH gradient** ($\Delta pH$) is itself a source of energy, since according to the laws of diffusion there must be a tendency for $H^+$ to move down its own concentration gradient and back into the mitochondrial matrix.

This combined electrochemical gradient is called the **proton-motive force** ($\Delta p$) for ATP synthesis:

$$\Delta p = \Delta\psi + \Delta pH$$

Clearly the more $H^+$ that is pumped out, the greater $\Delta p$ will be, and the more likely it is that ATP will be synthesised.

(iv) *The $H^+$ ions diffuse back into the matrix under their own electrochemical gradient through special membrane-bound ATPases. At these locally high concentrations, ATP synthesis occurs*

The identification of ATPases in the membrane strongly supports this postulate. Additional evidence is given in Q23.

> **Q23** Suppose the normal conditions are artificially reversed, i.e. suppose that the cytosol side of the inner membrane is made very alkaline compared with the matrix. What might you expect to happen to ATP?

*Other uses of the chemiosmotic pump*

Not all the energy in the $H^+$ electrochemical gradient is used to synthesise ATP. Some is used to drive the transport shuttle systems on the inner membrane (Fig. 4.9), or to pump $Ca^{2+}$ out of the cytosol and into storage (Chapter 3.2.3). In prokaryotes even greater use is made of the pump. Whilst bacteria never have mitochondria or chloroplasts, those capable of any form of oxidative respiration (Table 4.2) or photosynthesis have a chemiosmotic system attached to the plasma membrane. In these organisms, $\Delta p$ provides energy not only for the synthesis of ATP, but also for the active transport of nutrients such as amino acids, and even for the rotational movement of flagella.

**Study guide**

*Vocabulary*

Distinguish between:
oxidative phosphorylation and substrate-linked phosphorylation.

*Review Questions*

1 *The energetics of respiration*
Assume that one molecule of glucose is respired completely to $CO_2$ and $H_2O$:

$$6O_2 + C_6H_{12}O_6 \longrightarrow 6CO_2 + 6H_2O$$
$$(\Delta G^{o\prime} = -2870 \text{ kJ mole}^{-1})$$

By reference to Figs. 4.1 and 4.6:
(i)  Count how many ATPs are formed per glucose by substrate-linked phosphorylation.
(ii)  Count how many $NADH_2$s are formed per glucose in mitochondria. In oxidative phosphorylation each of these is worth three ATPs.
(iii)  Count how many $FADH_2$s are formed per glucose. In oxidative phosphorylation each of these is worth two ATPs.
(iv)  Count how many $NADH_2$s are formed per glucose during glycolysis. Moving this reducing power into the mitochondria by a shuttle system normally costs the equivalent of about one ATP, so score each of these $NADH_2$s as equivalent to two ATPs.
(v)  Calculate the number of ATPs produced by (i)–(iv). (By means of a table show precisely how you arrive at your answer.)
(vi)  Assume each ATP $\equiv$ 30 kJ mole$^{-1}$. What percentage of the total energy in glucose is captured as ATP? What happens to the rest?
(vii)  How many ATPs are synthesised when glycolysis goes to completion (forms lactate)? What is this as a percentage of the total amount of energy potentially available?

2 *The Pasteur effect*
Under aerobic conditions, yeast grows more vigorously, yet Pasteur noted that the amount of glucose consumed is significantly *less* than under anaerobic conditions. Explain these observations as fully as you can.

*Extension Questions*

(i) Explain why the terms glycolysis and anaerobic respiration are not completely interchangeable, and why oxidative respiration does not necessarily mean aerobic respiration.

(ii) The text states that there is no nett oxidation of glucose when glycolysis goes to completion, or during alcoholic fermentation, yet it refers to an oxidation step and $NADH_2$ production (Section 4.1(3). Explain this apparent inconsistency.

# Autotrophy

**SUMMARY**

External sources of energy, notably light, are used to generate reduced dinucleotides and ATP, which in turn drive the assimilation of $CO_2$, $NO_3^-$, and $SO_4^{2-}$ into organic materials. The conversion of light into chemical energy (ATP, $NADPH_2$) is a complex process, and in green plants involves two coupled photosystems. In the chloroplast, these photosystems are components of an ATP-synthesising chemiosmotic system similar to that found in mitochondria. There is only one pathway, the C3 pathway (syn.: Calvin cycle), capable of reducing $CO_2$ to the level of carbohydrate. However, in some plants this pathway may be supported by other pathways (C4, CAM) which are more efficient at trapping $CO_2$ from the atmosphere initially. C4 and CAM systems are especially important in the tropics, where they help to overcome the effects of photorespiration. Sections 5.1 to 5.1.2 are written in the form of revision exercises.

The living world's stockpile of organic materials, together with the potential energy in it, is continually eroded by processes such as respiration. This stockpile is replenished by the **autotrophs**: organisms capable of utilising external sources of energy for the manufacture of new organic materials from inorganic materials. Since respiration is exergonic, autotrophy is necessarily endergonic. To take glucose as an example, its synthesis requires a very considerable input of energy, at least equal to that released during its oxidation:

$$C_6H_{12}O_6 + 6O_2 \underset{\substack{\text{reduction}\\\text{(fixation)}}}{\overset{\substack{\text{(respiration)}\\\text{oxidation}}}{\rightleftharpoons}} 6CO_2 + 6H_2O$$

ATP

environmental energy

($\Delta G^{o'}$ (oxidation) $= -2870$ kJ mole$^{-1}$)

The reduction of $CO_2$ to carbohydrate $(CH_2O)_n$ has dominated our thinking about autotrophy because carbon forms the skeleton of all biochemicals, and carbohydrates characteristically accumulate during $CO_2$ fixation. Equally important for autotrophy, however, is the ability to reduce the oxidised forms of nitrogen and sulphur which occur in the environment (principally $NO_3^-$ and $SO_4^{2-}$) to the reduced forms found in amino acids and nucleotides. True autotrophy is only possible when reduction of all three of these inorganic substances occurs. Only then is an organism likely to be self-sufficient for its major organic requirements. Not all $CO_2$ fixers can utilise $NO_3^-$ (or $SO_4^{2-}$), however, and not all $NO_3^-$ (or $SO_4^{2-}$) reducers can fix $CO_2$ (Table 5.1). This implies that *at least* three distinct metabolic pathways are involved.

Table 5.1 *Strategies for nutrition.* Living organisms consist mainly of reduced forms of carbon, nitrogen, and sulphur, whereas the environment contains oxidised forms of these elements. Only a true autotroph is able to reduce all three elements to the levels found in protoplasm. (Reduced forms are in bold lettering.)

| Organism | Form in which element is used | | |
|---|---|---|---|
| | Carbon | Nitrogen | Sulphur |
| Green plants, some bacteria | $CO_2$ | $NO_3^-$ | $SO_4^{2-}$ |
| Animals | $CH_2O$ | $-NH_2$ | $-SH$ |
| Photosynthetic heterotrophs e.g. *Euglena* | $CO_2$ | $-NH_2$ | $-SH$ |
| Some fungi, e.g. *Neurospora* | $CH_2O$ | $NO_3^-$ | $SO_4^{2-}$ |

The environmental energy which is used by autotrophs for these reductive processes can vary. In all green plants and some bacteria, light energy is used (**photoautotrophy**). However, there is nothing magical about light, and some bacteria are able to use chemical energy from the environment instead (**chemoautotrophy**).

In oxidative respiration, hydrogen comes from organic acids via dinucleotides, provides energy for ATP synthesis and reduces an element (X) to $H_2X$. In what may be called the primary autotrophic event this sequence is essentially reversed. Thus, in green plants, light energy is used to drive hydrogen atoms from $H_2O$. The hydrogen is then used to synthesise ATP and to reduce dinucleotides. In the secondary autotrophic events, ATP and reduced dinucleotides are then used (directly or indirectly) to reduce $CO_2$, $NO_3^-$ and $SO_4^{2-}$ (Fig. 5.1).

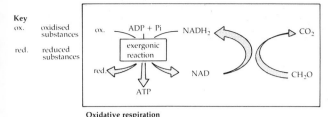

Key
ox.  oxidised substances

red.  reduced substances

**Oxidative respiration**

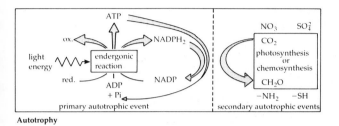

**Autotrophy**

Fig. 5.1 *Oxidative respiration and autotrophy.* In most organisms 'red.' = water, and 'ox.' = oxygen, but in some organisms, especially bacteria, these abbreviations may refer to other substances such as $H_2S$ and S. Light energy is *not* required in chemoautotrophs such as nitrifying bacteria, since in these organisms oxidation of reduced substances (such as $NH_4^+$ to $NO_3^-$) is exergonic. (How would you amend the lower diagram to make it applicable to chemoautotrophs?)

## 5.1 PHOTOSYNTHESIS IN EUKARYOTES: BASIC CONCEPTS

The utilisation of light energy by eukaryotic green plants for the assimilation of $CO_2$ into carbohydrate is without doubt quantitatively the most important of all autotrophic processes. About 0.2% of the Earth's total carbon—75 billion ($75 \times 10^9$) tonnes—is fixed by photosynthesis per year. Some of the basic features of photosynthesis are reviewed below (Q1–Q13).

*Requirements*

Photosynthesis is a complex process with a number of specific requirements.

**Q1** State the essential requirements for photosynthesis (other than a supply of ADP, Pi and NADP).

**Q2** How could you demonstrate a requirement for $CO_2$, light and chlorophyll?

It is not easy to demonstrate a requirement for water, since water is used in many processes other than photosynthesis. A water shortage may therefore result in fatal consequences for a variety of reasons and long before photosynthesis is affected.
Its involvement in the process can be convincingly demonstrated, however, using labelled water, $H_2^{18}O$ (Q20).

*Products*

Regarding the products of photosynthesis, experiments demonstrating oxygen evolution date back as far as Priestley (1772).

**Q3** What did Priestley's experiment involve?

**Q4** A more sophisticated experiment was carried out by Engelmann in 1883 (Fig. 5.2). What conclusions may be drawn from this?

Readers may recall from elementary science courses the starch–iodine test of leaf samples as evidence for photosynthesis. This test is, however, fraught with difficulties. In the first place, some plants, such as lilies, never form starch. Although they form other polysaccharides their leaves do not turn blue-black with iodine. Secondly, starch production, even if it does occur, is biochemically several steps removed from photosynthesis itself. Thirdly, starch synthesis can be induced in the dark, provided that the appropriate requirements are met. (Starch is never transported in plants, yet potatoes, which never see daylight, are full of it!) In short, in order to demonstrate photosynthesis or to measure its rate, we need to find a better alternative than testing leaves for starch.

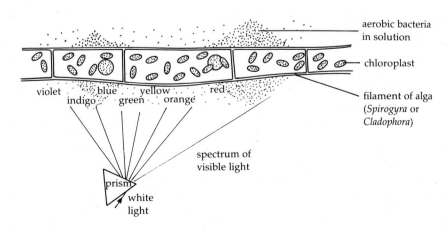

Fig. 5.2 *Engelmann's experiment*

### 5.1.1 Measuring the rate

In principle anything consumed or produced can be used to measure the rate of photosynthesis. In practice the choice is more restricted. For many purposes the rate is most conveniently measured in terms of $CO_2$ uptake or $O_2$ evolution, using aquatic plants (especially unicellular algae).

**Q5** Suggest two advantages of using cultures of unicellular algae over potted plants to determine rates of photosynthesis.

**Q6** In what units can $CO_2$ uptake or $O_2$ evolution be conveniently expressed?

**Q7** What practical or theoretical problems make it more difficult to measure the rate using, say, $H_2O$ consumption, carbohydrate production or light uptake?

Respiration introduces a complication because it is occurring simultaneously and is consuming $O_2$ and producing $CO_2$. A record of, say, the number of $mm^3$ $O_2$ produced $g^{-1}$ algae $min^{-1}$ is therefore only a measure of the *nett* result of:

$O_2$ *produced* by photosynthesis   *minus*   $O_2$ *consumed* by respiration

The amount of respiration can be estimated by calculating $O_2$ consumption in the dark, when there is no photosynthesis. If the value obtained is added to the value for $O_2$ production in the light a better estimate of the real rate of photosynthesis may be obtained. There is, however, an enormous assumption being made in this line of argument.

**Q8** What assumption is being made?

In Section 5.5 it will be shown that $O_2$ uptake and $CO_2$ production can be stimulated by light. This effect, which mimics respiration in the gases exchanged, is called **photorespiration**. Hence:

$$\frac{\text{true}}{\text{rate}_{ph}} = \frac{\text{apparent}}{\text{rate}_{ph}} + \frac{\text{allowance for}}{\text{respiration}} + \frac{\text{allowance for}}{\text{photo-}}_{\text{respiration}}$$

    ($O_2$ produced) ($O_2$ consumed) ($O_2$ consumed)

Measuring the rate of photorespiration is difficult. In many of the experiments described below it is ignored; so is respiration. In such cases 'rate of photosynthesis' really means '*apparent* rate'.

### 5.1.2 Limiting factors and the compensation point

Blackmann (1905) proposed that the rate of photosynthesis was limited by whatever requirement was in shortest supply (the **Law of Limiting Factors**). Figure 5.3 demonstrates this principle.

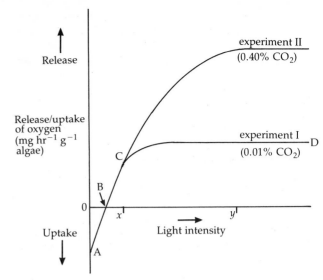

Fig. 5.3 *The effects of light and $CO_2$ concentration on the rate of photosynthesis*

**Q9** In the region A–B, is there a nett uptake or a nett release of $CO_2$? How do you explain this?

**Q10** Point B is called the **compensation point**. At this point $CO_2$ production due to respiration just balances $CO_2$ uptake due to photosynthesis. Comment on the ecological significance of the compensation point in (i) a forest, (ii) a river and (iii) a cave.

**Q11** What factor is limiting the rate between (i) A and C, and (ii) C and D?

**Q12** In experiment I, an increase in light intensity between $x$ and $y$ had no effect on the rate, whereas it did in experiment II. Explain.

**Q13** How might a further increase in the rate be achieved at light intensity $y$ over that achieved in experiments I and II?

### 5.1.3 Evidence for separate light and dark reactions

A feature of photosynthesis which intrigued plant physiologists for a long time was the dual effect of both light and heat on the reaction. Chemical reactions fall into two categories: **thermochemical** and **photochemical**. The former are the only kind that have been mentioned so far in the text. They are temperature dependent and unaffected by light. Conversely, photochemical reactions are light dependent and temperature independent. So, in photosynthesis, are there separate photochemical and thermochemical reactions? Alternatively, does photosynthesis involve a unique 'photothermochemical reaction'? Emmerson and Arnold (1932) designed an experiment to distinguish the two possibilities. In the experiment batches of *Chlorella* were kept at either 1 °C or 25 °C. Each batch was divided into four samples. The four pairs of samples were then given *precisely the same total quantity of illumination but in different ways*, and the amount of photosynthesis was measured.

(i)  One pair of samples (1 °C, 25 °C) were given all their light at once. In the remaining pairs, light was given in flashes of $10^{-5}$ s each, with varying dark intervals between each flash:
(ii)  1/10 s between flashes;
(iii)  2/10 s between flashes;
(iv)  4/10 s between flashes.

> **Q14** If photosynthesis consisted of just a single 'photothermochemical reaction', should it make any difference whether the light is given continuously or in flashes?

Emerson's results are shown in Fig. 5.4. A and B correspond to the amounts of photosynthesis in continuous light, at 1 °C and 25 °C. Since very bright light was used it is not surprising that Emmerson observed that B was considerably greater than A. At 25 °C the lengthening dark interval had no significant effect (C, D, E). However, at 1 °C the amount of photosynthesis *per given amount of light* increased as the length of the dark intervals increased (F, G, H). They concluded that photosynthesis consisted of separate light and dark reactions (Q15, Q16).

> **Q15** Recall: photochemical reactions are light dependent and temperature independent; thermochemical reactions are temperature dependent and light independent.
> (i)  At 1 °C, which type of reaction would limit the rate of photosynthesis (given adequate illumination)?
> (ii)  Can a light reaction occur when the light is off (between flashes)? Why does the sample at 1 °C 'catch up' with the 25 °C sample when there is a long enough dark interval between flashes?
>
> **Q16** Draw a series of histograms to show what results might have been obtained if there was only one type of 'photothermochemical reaction'.

> **Q17** The batch at 25 °C behaved like the batch at 1 °C if it was treated with dilute hydrogen cyanide solution. Explain.

The term 'dark reaction', which is normally applied to the thermochemical reaction, is a little confusing: it simply means that light is irrelevant. Normally it takes place simultaneously with the light (photochemical) reaction during daylight hours, though it *can* occur in the dark if appropriate materials are provided.

## 5.2 THE LIGHT REACTION

Showing that a distinct light-dependent reaction exists was relatively straightforward. Working out what it does and how it works has been much more difficult. Some details remain unsolved, but there are a number of features about which we can be certain:
(i)  *The light reaction evolves oxygen (but it does not consume $CO_2$; neither does it produce carbohydrate)*
Using isolated chloroplasts and an electron acceptor ($CO_2$ substitute), Hill (1937) showed that oxygen could be produced when the chloroplasts were illuminated. In the experiment no $CO_2$ was consumed, and no carbohydrate was produced.
(ii)  *The oxygen evolved during photosynthesis comes from water, not $CO_2$*
Ruben and Kamen (1941) demonstrated that $^{18}O_2$ was evolved if photosynthesis was allowed to take place in the presence of labelled water, $H_2^{18}O$.

> **Q18** What instrument is used to detect $^{18}O_2$?
>
> **Q19** Explain Ruben's result, illustrating your answer by means of a balanced chemical equation which shows glucose as the photosynthetic product.
>
> **Q20** What control experiment is necessary to confirm the conclusion that $O_2$ comes from water, not $CO_2$?

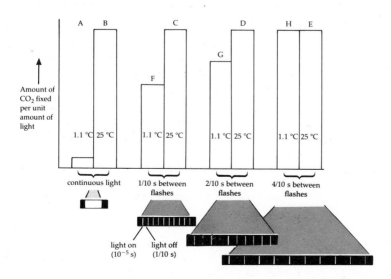

Fig. 5.4 *Emmerson's flashing-light experiments.* In all cases very high intensity white light was used. The flash duration was $10^{-5}$ s in experiments (ii) − (iv). The total quantity of light delivered was the same in all experiments.

(iii) *The hydrogen component of $H_2O$ is used to generate $NADPH_2$ and ATP*

The synthesis of $NADPH_2$ and ATP is the primary function of the light reaction. Arnon *et al.* (1954) demonstrated that in isolated illuminated chloroplasts the evolution of oxygen from water was linked both to the reduction of NADP and to ATP synthesis.

$$H_2O + NADP \longrightarrow NADPH_2 + \tfrac{1}{2}O_2$$
$$\overset{ADP}{\underset{+ Pi}{}} \overset{\frown}{\quad} ATP$$

(non-cyclic photophosphorylation)

Since ATP synthesis in this case appeared to depend upon a light-driven linear flow of electrons from $H_2O$ to NADP, it was called **non-cyclic photophosphorylation**. Three years previously the same group had shown that under different conditions ATP synthesis could occur *independently* of NADP utilisation and oxygen evolution. This phenomenon, **cyclic photophosphorylation**, couples ATP synthesis to an endless recycling of a light-driven electron from chlorophyll.

$$ADP + Pi \longrightarrow ATP \quad \text{(no NADP or } H_2O$$
$$\text{consumed; no } O_2 \text{ evolved;}$$
$$\text{no } NADPH_2 \text{ produced)}$$

(cyclic photophosphorylation)

Arnon's discoveries were of immense significance because they demonstrated the precise function of the light reaction. By 1954 it was already known that both ATP and $NADPH_2$ were consumed during the fixation of $CO_2$ in the dark reaction. Hence the lower drawing in Fig. 5.1 could by 1954 be relabelled 'light reaction' and 'dark reaction' (primary and secondary autotrophic events respectively).

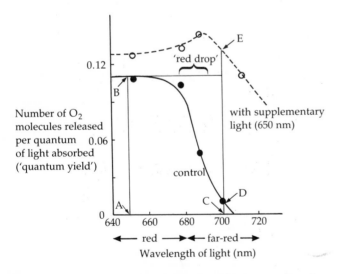

Fig. 5.5 *The red drop and the Emmerson enhancement effect.* It can be shown that although chlorophyll strongly absorbs light between 680 and 700 nm, the amount of photosynthesis per unit of light absorbed diminishes greatly at these wavelengths, i.e. photosynthesis becomes inefficient. This effect — **the red drop** — disappears if the longer wavelengths are supplemented by light <650 nm. Moreover, the total amount of photosynthesis using the mixture of wavelengths (C-E) is greater than the sum of that obtained by using either independently (C-D + A-B). This is called the **enhancement effect**.

(iv) *The light reaction actually consists of two cooperative photosystems*

Emmerson (1957) measured the amounts of photosynthesis for given amounts of light at 650 nm and 700 nm. He then illuminated the same plants with the same *quantity* of light, but provided *both* wavelengths at the *same time*. The result was conclusive. The amount of photosynthesis when the two wavelengths were provided simultaneously greatly exceeded the sum of the amounts when the same wavelengths were provided separately. This was called the **Emmerson enhancement effect** (Fig. 5.5: distance CE > (CD) + (AB). Emmerson concluded that the light reaction actually consisted of *two cooperative photosystems* which operate at slightly different wavelengths.

Since different chemicals absorb different wavelengths (see Appendix), Emmerson's conclusion also predicts the existence of subtle differences in the pigments between the two photosystems. This has since been confirmed (Section 5.2.1).

(v) *The two photosystems (called PSI and PSII) are linked in series during $NADPH_2$ production*

'In series' means 'one after the other':

$$\left.\begin{array}{c}H_2O \\ \tfrac{1}{2}O_2\end{array}\right\} \xrightarrow{\quad PSII \quad} \xrightarrow{\quad PSI \quad} \left\{\begin{array}{c}NADP/ADP + Pi \\ NADPH_2/ATP\end{array}\right.$$

Some evidence supporting this proposition is given in the box below.

---

Before attempting Q21 and Q22, readers may find it helpful to consult the Appendix, and also the note following Q22.

**Q21** What does the reciprocal of the vertical ($y$) axis in Fig. 5.5 tell us? This reciprocal is called the **quantum efficiency**. What is the average quantum efficiency (below 670 nm)? Round off to the nearest whole number (according to the laws of physics, quanta don't exist in fractions!).

**Q22** While Emmerson's work suggests that 8 quanta are needed to release one $O_2$ molecule, this seems to conflict with Ruben's work (para (ii)). Replace $A$, $B$, $C$ and $D$ with appropriate whole numbers so that *all* the $O_2$ evolved comes from water:

$$AH_2O \underset{COH^-}{\overset{BH^+ + De}{\diagdown}}$$
$$BH^+ + De \xrightarrow{\quad CO_2 \quad} (CH_2O) + H_2O$$
$$COH^- \longrightarrow O_2 + 2H_2O$$

A basic feature of photochemistry is that in a *single* reaction *one* electron is excited by *one* quantum: there is a fundamental 1:1 relationship. It follows that during the light reaction *four* electrons (Q22) must be excited by *two sequential* reactions if the quantum efficiency is about 8 (Q21).

4 electrons × 2 coupled photochemical
(*observed*: Q22) reactions (*deduced*)

≡ 8 quanta
(*observed*: Q21)

**(vi)** *The linked photosystems are connected by an electron transport chain consisting of redox agents*
The redox agents are mostly quinones and cytochromes similar to those found in the electron transport chain of mitochondria. The simple sequence of photosystems illustrated in (v) does not properly take into account the ability of the chloroplast components to receive or pass on electrons. Neither does it explain the effect of light on chlorophyll. In an inspired article, Hill and Bendall (1960) accounted for both phenomena, proposing what has been called the Z-scheme. See if you can deduce it from Q23–Q25.

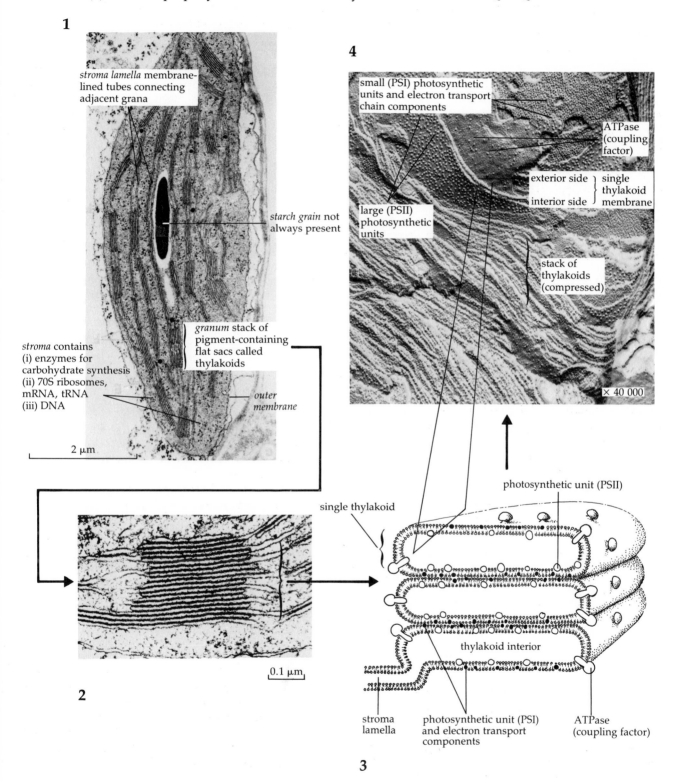

Fig. 5.6 *Structure and function in chloroplasts.* 1, sectioned whole chloroplast; 2, granum, detail; 3, thylakoid structure; 4, freeze-fractured thylakoid membrane. Whilst the thylakoids are the **transducers** which convert light energy to chemical energy, the stroma is the site of biochemical synthesis. The latter contains not only the enzymes for the fixation of $CO_2$ to carbohydrate, but also those for the reduction of nitrate and sulphate, and for the synthesis of fatty acids. All these processes depend on ATP and reducing power provided by the light reaction. (In animals nitrate and sulphate cannot be reduced, and fatty acid synthesis occurs in the cytosol.)

Recall (Chapter 1.3.2):
(i) *oxidation* = electron (or hydrogen) loss; *reduction* = electron (or hydrogen) gain;
(ii) moving from negative to positive on the redox scale (Fig. 5.Q23) releases energy; moving 'up' the scale requires energy.

Hence in the light reaction of photosynthesis, light energy is used to drive hydrogen from water ($+0.82$ V) to NADP ($-0.34$ V) (Fig. 5.Q23).

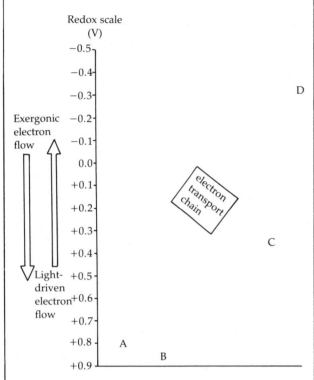

Fig. 5.Q23 *Deducing the Z-scheme.* Your answer to Q23–Q25 should produce a simplified version of the scheme. ADP + Pi are converted to ATP as electrons flow through the electron transport chain (add this to your diagram).

Sketch the axes of Fig. 5.Q23 and add the electron transport chain as shown.

**Q23** A–D correspond to NADP/NADPH$_2$, unenergised photosystems I and II, and water. Use Table 1.1 to decide which letter applies to which substance. Substitute the name of the substance for the letter on your sketch.

**Q24** An input of a quantum into a PS excites an electron making the PS 0.9 V more negative. Draw vertical lines from PSI and PSII corresponding to $-0.9$ V. Put double arrowheads (↟) on the lines to show that light energy is required.

**Q25** Remembering that electrons flow 'downwards' in exergonic reactions, show the path of electrons (or hydrogen) from H$_2$O to NADP by connecting all the components with one appropriate line. Indicate exergonic steps by a single downward-pointing arrow (↓).

(vii) *Photosystems I and II must both operate to generate NADPH$_2$ and ATP*

If only photosystem I is active, only ATP is synthesised. These conclusions may be drawn from experiments which involve the use of drugs to inactivate one or other system and from mutants in which one or other system is missing.

### 5.2.1 Chloroplast organisation

The essential features of the chloroplast have been worked out using both electron microscopy and biochemical analysis (Fig. 5.6).

*The photosynthetic pigments*

There are several pigments associated with photosynthesis. Each pigment absorbs light at particular wavelengths (Table 5.2).

Table 5.2 *Photosynthetic pigments.* The absorption spectra for the pigments in *organic solvents* are unique and characteristic. However, the real biological significance of such spectra is uncertain, because they probably change when they are associated with proteins in the thylakoid membrane. Detailed absorption spectra have therefore been omitted from the table deliberately.

| Pigment | Occurrence and comments |
|---|---|
| *Chlorophylls* | Absorb mainly blue and red |
| Chlorophyll *a* (bluish green) | All photosynthetic eukaryotes. More abundant in PSI |
| P700 | A species of chlorophyll *a* which absorbs at 700 nm and is associated with PSI |
| P680 | A species of chlorophyll *a* which absorbs at 680 nm and is associated with PSII |
| Chlorophyll *b* (yellow-green) | All green plants. More common in PSII |
| Others (*c, d*) | Various non-green algae |
| *Carotenoids* | Absorb mainly blue-green |
| β-carotene (orange-yellow) | All photosynthetic eukaryotes except some algae. Probably passes energy on to chlorophyll; may also prevent photo-destruction of chlorophyll at high light intensities |
| Others (e.g. fucoxanthin, α-carotene) | Various algae and to some extent in green plants |
| *Phycobilins* | Absorb mainly blue-green |
| Phycocyanin | Blue-green 'algae' (*Cyanobacteria*) |
| Phycoerythrin | Red algae. Traps shorter wavelengths than chlorophyll, i.e. those wavelengths which predominate in the deeper waters where these algae are found. |

The main pigments in eukaryotes are the chlorophylls. In higher green plants these fall into two categories, **chlorophyll *a*** and **chlorophyll *b***. The accessory pigments such as **carotenoids** and **xanthophylls** pass on some or all of the energy they absorb to the chlorophylls, so that the **action spectrum** for photosynthesis is a composite of the various **absorption spectra** of all the different pigments (Appendix Fig. A.1).

*Organisation of the photosystems in the thylakoid*

PSI contains a special species of chlorophyll *a* called $P_{700}$ (absorbs at 700 nm) while PSII contains a different species called $P_{680}$ (absorbs at 680 nm). Structurally each photosystem seems to be surrounded by an assembly of about 250 chlorophyll *a* and *b* molecules (the **antenna complex**) which trap light energy and funnel it into $P_{700}$ or $P_{680}$ (the **reaction centres**). The accessory pigments, which tend to absorb at shorter wavelengths, are loosely associated with both systems, especially with PSII.

The photosynthetic pigments are embedded entirely in the **thylakoid membranes**, predominantly at the regions where adjacent thylakoids meet. It is in these membranes therefore that the light reaction occurs. As Fig. 5.7 indicates, the distribution of the photosystems and the electron transport chain components in the membrane is not symmetrical. This asymmetry is essential for the operation of the light reaction.

### 5.2.2 Chemiosmosis in chloroplasts

(i) *Non-cyclic photophosphorylation*
In non-cyclic photophosphorylation a linear flow of light-energised electrons is driven from water inside the thylakoid sac to NADP outside the sac. $H^+$ accumulates on the inside, partly originating from the $H_2O$ which is consumed during the light reaction, and partly from $H^+$ ions in the cytosol. The latter are pumped into the thylakoid during intermediate reactions which occur as electrons flow along the asymmetrically distributed electron transport chain. The thylakoid membrane, like the inner mitochondrial membrane, is impermeable to $H^+$ except at regions where **coupling factors** (*ATPases*) are embedded. At these points, as in mitochondria, $H^+$ diffuses down its own electrochemical gradient and synthesises ATP by a reverse hydrolysis as previously described (Chapter 4.4).

(ii) *Cyclic photophosphorylation*
PSI by itself can synthesise ATP (but not $NADPH_2$) by recycling an electron as shown by the dotted line in Fig. 5.7.

### 5.3 THE DARK REACTIONS

There is only one pathway, the **C3** or **Calvin–Benson pathway**, capable of reducing $CO_2$ to the level of carbohydrate ($CH_2O$). The enzymes and other materials for this are found exclusively in the **stroma** of the chloroplast, and it is undoubtedly here that carbohydrate synthesis occurs. One of two other patterns of $CO_2$ fixation may be present in some plants in addition to the C3 pathway. These are the **C4** and **CAM** pathways. If present, they occur in the **cytosol**, not in the chloroplasts. Their function is the initial fixation of $CO_2$ into organic material in circumstances where the C3 pathway alone is likely to be inefficient.

### 5.3.1 The C3 (Calvin–Benson) pathway

Calvin and Benson (1947) elucidated the C3 pathway using $^{14}CO_2$ as a tracer (Fig. 5.8 and Q26–Q27).

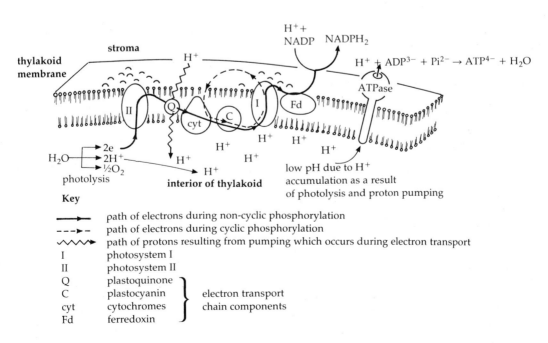

Fig. 5.7 *The light reaction in the chloroplast*

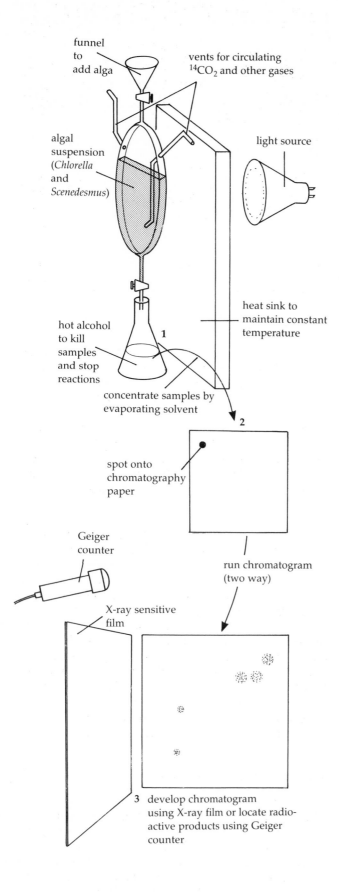

Fig. 5.8 *Calvin's technique for the elucidation of the dark reaction*

**Q26** The chromatograms below show which compounds became radioactive 5 s and 15 s after the start of the experiment. What do they suggest is the first formed compound? How many carbons does this compound contain?

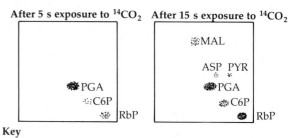

**Key**

| | | | |
|---|---|---|---|
| PGA | phosphoglyceric acid | MAL | malate |
| C6P | glucose and other hexose phosphates | ASP | aspartate |
| RbP | ribulose bisphosphate | PYR | pyruvate |

Fig. 5.Q26 $^{14}C$ *positions after 5 s and 15 s exposure to* $^{14}CO_2$

**Q27** Among the substances formed later is a C5 sugar, **ribulose bisphosphate**. The concentration of this and PGA is dramatically altered when the $^{14}CO_2$ supply is shut down (Fig. 5. Q27). Explain the changes.

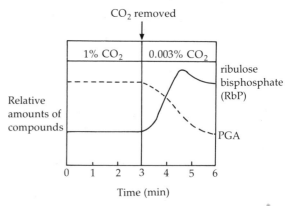

Fig. 5.Q27 *Changes in PGA and RbP concentrations following removal of* $CO_2$

The first step is **carboxylation**. *Ribulose bisphosphate carboxylase*, the enzyme which combines $CO_2$ with RbP to form two PGA molecules, accounts for over 50% of all the protein in the stroma. Indeed, *RbP carboxylase* can claim the title of being the world's most abundant protein.

To convert an acid such as PGA into a sugar (aldehyde or ketone), it must be **reduced**. This is the next step, and it is accomplished using $NADPH_2$ from the light reaction. Although most of the energy to drive the reaction comes from the oxidation of $NADPH_2$ itself, some is provided by ATP (Q30).

$$NADPH_2 + \text{phosphoglyceric acid (PGA)} \longrightarrow$$

$$ATP \longrightarrow ADP + Pi$$

glyceraldehyde-3-phosphate (GAP) + NADP

Fig. 5.9 *Spartina townsendii*. This relative newcomer to the British flora has successfully invaded several coastal communities. The leaf shows typical Kranz anatomy (right). Two families of monocots and eight families of dicots exhibit C4 metabolism, suggesting that the pathway may have evolved independently on more than one occasion (i.e. it is polyphyletic).

---

> **Q28** In the reduction phase how many ATPs and NADPH$_2$s are used for each CO$_2$ incorporated?

The third step involves the conversion of GAP (C3) into a variety of C4 to C7 sugars, a proportion of which ultimately form ribulose phosphate (RuP). This is then converted back to ribulose bisphosphate by ATP (Fig. 5.11).

$$\text{ATP} + \text{RuP} \longrightarrow \text{RbP} + \text{ADP}$$

> **Q29** In this final phosphorylation stage how many ATPs are needed to form a *single* RbP molecule?
>
> **Q30** Taking Q28 and Q29 together, what is the total NADPH$_2$ and ATP requirement for each CO$_2$ molecule? Given $\Delta G^{o\prime}$, ATP hydrolysis $= -30$ kJ mole$^{-1}$, and NADPH$_2$ oxidation $= -220$ kJ mole$^{-1}$, what provides most of the energy to drive CO$_2$ fixation?
> (Quantify your answer.)

Most of the remaining C3–C7 sugars are used for synthetic purposes. All organic carbon originates from the Calvin cycle, and consequently lipids, amino acids and nucleotides will be synthesised in addition to carbohydrate. Some of the intermediates may also directly enter the respiratory pathways (PGA is an intermediate of glycolysis, and is readily transported through the chloroplast membrane).

> **Q31** Assume, for simplicity, that six CO$_2$ molecules are channelled directly into glucose synthesis. What is the energy requirement per glucose synthesised? (ATP/NADPH$_2$ is only required at the steps described above.) Given that glucose oxidation yields 2870 kJ mole$^{-1}$, how efficient is the Calvin cycle?

### 5.3.2 The C4 (Hatch and Slack) pathway

As an enzyme, *RbP carboxylase* leaves much to be desired. It is competitively inhibited by O$_2$, leading to **photorespiration** (Section 5.5), and it becomes virtually inactive if the CO$_2$ concentration drops to about 30 ppm (1/10 of its atmospheric value). In hot, sunny or salty conditions, water stress can induce stomatal closure for several hours even during the day. The CO$_2$ concentration may then drop to very low levels indeed. Many tropical plants such as sugar cane and maize, and even some coastal temperate plants (Fig. 5.9), have solved the problem by developing a CO$_2$-concentrating system.

All these plants show **Kranz anatomy**, which means that the vascular bundles in the leaves are surrounded by a ring of **bundle sheath cells**. The chloroplasts of these cells have two distinctive features. First, they often have few grana, and in some cases none at all (in which case they are described as **agranal**). Secondly, they contain virtually all the plant's supply of *RbP carboxylase*. The chloroplasts of the mesophyll cells always look 'normal', but lack *RbP carboxylase*. However, the *cytosol* of the mesophyll cells contains another CO$_2$-fixing enzyme, *PEP carboxylase* (*phosphoenolpyruvate carboxylase*) (Figs. 5.10 and 5.11).

*PEP carboxylase* can fix CO$_2$ at extremely low CO$_2$ levels (and shows no obvious inhibition by O$_2$):

$$\text{PEP (C3)} + \text{CO}_2 \xrightarrow[\text{in mesophyll cells}]{\textit{PEP carboxylase}} \text{oxaloacetate (C4)}$$

The oxaloacetate produced is reduced to malate (or other C4 acids) and moved to the bundle sheath cells. Here the C4 acids are **decarboxylated**, with the nett result that CO$_2$ initially captured over a large volume of leaf space is released into the relatively small volume of the bundle sheath. Consequently the local CO$_2$ concentration in these cells is very high. The CO$_2$ is then refixed by *RbP carboxylase* into the Calvin cycle, whilst pyruvate is re-exported to the mesophyll cells

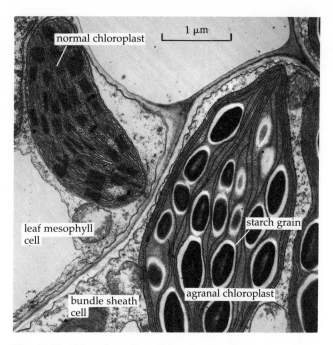

Fig. 5.10 *Chloroplast dimorphism in C4 plants.* Not all C4 bundle sheath chloroplasts lack grana. It seems to be a rather variable characteristic, more commonly associated with advanced genera such as the grasses and maize.

where it is turned back into PEP (Fig. 5.11). The refixation by *RbP carboxylase* is necessary because there is no mechanism for directly reducing the $CO_2$ trapped in malate (as —COOH) to sugar (—CHO).

There is a price to pay for capturing atmospheric $CO_2$ at very low concentrations. ATP is needed to convert pyruvate to PEP, and perhaps also for the transport of organic acids to and from the mesophyll cells:

$$\text{pyruvate} + \text{Pi} + \text{ATP} \xrightarrow[\substack{\text{in}\\\text{mesophyll}\\\text{cells}}]{\substack{\text{pyruvate}\\\text{dikinase}}} \text{PEP} + \text{AMP} + \text{PP}$$

In the tropics, ATP is abundantly available as a result of the light reaction. The higher energy cost of this double fixation (C4 followed by C3) is perhaps a small price for a plant to pay so that it can use $CO_2$ at low concentrations. However, C4 plants have not generally superseded C3 plants in temperate regions. This suggests that as temperature and light availability diminish the 'energy cost' becomes too high, and the advantage C4 plants possess in the tropics is lost.

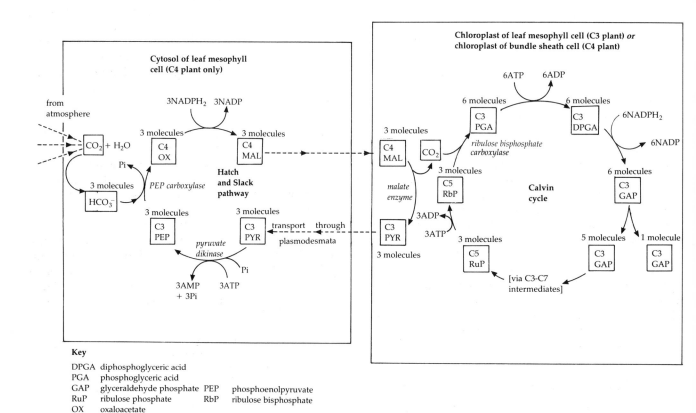

Key

| | |
|---|---|
| DPGA | diphosphoglyceric acid |
| PGA | phosphoglyceric acid |
| GAP | glyceraldehyde phosphate |
| RuP | ribulose phosphate |
| OX | oxaloacetate |
| MAL | malate |
| PYR | pyruvate |

| | |
|---|---|
| PEP | phosphoenolpyruvate |
| RbP | ribulose bisphosphate |

Fig. 5.11 *The C3 and C4 pathways.* In C3 plants only the Calvin cycle operates, trapping atmospheric $CO_2$ in the leaf mesophyll cells (bundle sheath cells being absent). In C4 plants the Hatch and Slack cycle operates in the leaf mesophyll cells, exporting 'trapped $CO_2$' via a carbon carrier to the bundle sheath cells, where the $CO_2$ is refixed by the Calvin cycle. (Key enzymes are shown in italics.)

Fig. 5.12 *CAM plants*. Crassulacean acid metabolism is now known to occur in several families of flowering plants in addition to the *Crassulaceae*, including various genera of the *Euphorbiaceae, Liliaceae, Orchidaceae* and *Cactaceae*.

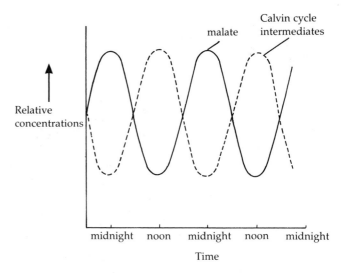

Fig. 5.13 *Diurnal fluctuations in malate and Calvin cycle intermediates in CAM plants*

### 5.3.3 CAM plants

C4 plants create locally high $CO_2$ concentrations by effectively moving $CO_2$ from all over the leaf to the bundle sheath cells. CAM plants maintain higher-than-atmospheric levels by altering the time of fixation. CAM stands for **crassulacean acid metabolism**, after the angiosperm family *Crassulaceae* in which it was first discovered.

These plants show a reversed stomatal rhythm. In other words the stomata open at night and close during the day. By opening when it is cool and when the atmosphere is generally more humid, water loss is considerably reduced. Labelling experiments show that $^{14}CO_2$ is fixed into malate in the mesophyll cells at night, when the stomata are open. The acid is then stored overnight in the sap vacuole. During the day when the stomata are closed it is re-exported to the cytoplasm and broken down to pyruvate, so releasing $CO_2$ (Fig. 5.13). Biochemically C4 and CAM plants are therefore rather similar, although the different strategies they have adopted have resulted in some marked biological differences (Table 5.3).

### 5.4 NITROGEN AND SULPHUR ASSIMILATION

As previously emphasised, a true autotroph is able to utilise inorganic sources of nitrogen and sulphur as well as carbon, and so become self-sufficient for all its nutritional requirements. Qualitatively nitrogen and sulphur assimilation are just as important to the continuance of life as carbon assimilation, even though the rate at which they are assimilated (about 0.3 billion tonnes $yr^{-1}$) is less than 1% of the rate of carbon assimilation. In organic compounds, both elements exist mainly in reduced forms ($-NH$, $-NH_2$, $-SH$, $=S$) whilst in the environment they exhibit various degrees of oxidation, depending upon the local availability of oxygen and the action of micro-organisms (Table 5.4). The varied patterns of nutrition exhibited by the latter are major ingredients of the familiar nitrogen and sulphur cycles. They are more adequately discussed in *Microbes and Biotechnology* in this series, and the remainder of this section will be restricted to the assimilation of nitrogen and sulphur in green plants.

### 5.4.1 Nitrogen

Green plants have no mechanism for utilising the most abundant form of nitrogen, $N_2(g)$, except indirectly as cosymbionts of $N_2$-fixing prokaryotes such as *Rhizobium*. Whilst labelling experiments demonstrate that in principle plants are able to use a variety of salts such as $NO_3^-$, $NO_2^-$ and $NH_4^+$, the last two are exceedingly toxic. In practice, the form in which nitrogen is principally absorbed into roots is as $NO_3^-$. After it has been taken into cells by active transport it is progressively reduced by *nitrate* and *nitrite reductase* systems. Both systems are found in most plant tissues.

$$NO_3^- \xrightarrow[\text{in cytosol of most cells}]{\textit{nitrate reductase}} NO_2^-$$

$$NADH_2 \curvearrowright NAD + H_2O$$

$$NO_2^- \xrightarrow[\text{cytosol or chloroplasts}]{\textit{nitrite reductase}} NH_4^+$$

$$\begin{array}{cc} NADH_2 & NAD \\ \text{or} & \text{or} \quad + 2H_2O \\ Fd_{red} & Fd_{ox} \end{array}$$

Table 5.3 *Summary of C3, C4, and CAM plants*

| | C3 plants | C4 plants | CAM plants |
|---|---|---|---|
| Examples | Sunflower; sugar beet | Maize; sugar cane | Cacti; succulents |
| Habitat | Temperate regions | Mostly tropical; sometimes temperate | Arid regions |
| Anatomy | (i) No obvious bundle sheath cells<br>(ii) Monomorphic chloroplasts | (i) Bundle sheath cells<br>(ii) Dimorphic chloroplasts | (i) No bundle sheath cells<br>(ii) Monomorphic chloroplasts |
| $CO_2$ fixation:<br>(i) enzymes | Ribulose bisphosphate carboxylase | PEP carboxylase then ribulose bisphosphate carboxylase | As for C4 |
| (ii) compensation point for $CO_2$ | 1/10 atmospheric concentration | < 1/30 atmospheric concentration | As for C4 at night, similar to C3 during the day |
| Stomata | Open by day | Open by day | Open by night |
| Transpiration ratio (g $H_2O$ lost/g $CO_2$ fixed) | 500–1000 | 250–350 | 50–55 |
| Optimum temperature (°C) | 15–25 | 30–38 | 35–40 |
| Saturating light intensity | 1/5 full sunlight | > full sunlight | As for C4 |
| Losses due to photorespiration? (Section 5.5) | Yes | No | As for C4 |
| Maximum net gain (mg $CO_2$ fixed $dm^{-2}$ leaf $hr^{-1}$) | 15–40 | 40–80 | 1–10 |
| Net primary productivity (tonnes $hectare^{-1}$ $yr^{-1}$) | 20–25 | 22–55 | (Extreme variations in available data) |

Table 5.4 *The action of some prokaryotes on nitrogen and sulphur and their oxides*

| Category | Reactions | Examples |
|---|---|---|
| *Chemoautotrophs* | | |
| Energy-yielding oxidation of reduced minerals present in the environment is used to generate ATP and reduced dinucleotides (especially $NADH_2$). An alternative to the light reaction of photosynthesis | $S + 1\frac{1}{2}O_2 + H_2O \rightarrow SO_4^{2-}$<br>$NH_3 + 1\frac{1}{2}O_2 \rightarrow NO_2^- + H_2O + H^+$<br>$NO_2^- + \frac{1}{2}O_2 \rightarrow NO_3^-$<br>$H_2 + \frac{1}{2}O_2 \rightarrow H_2O$ | *Thiobacillus*<br>*Nitrosomonas*<br>*Nitrobacter*<br>*Hydrogenomonas* |
| *Anaerobic heterotrophs* | | |
| Oxidative respiration of glucose is achieved in anaerobic conditions by substituting oxidised mineral salts onto which $NADH_2$ can unload hydrogen | Glucose + $4.8HNO_3 \rightarrow$<br>$6CO_2 + 2.4N_2 + 8.4H_2O$<br><br>Glucose + $3H_2SO_4 \rightarrow$<br>$6CO_2 + 3H_2S + 6H_2O$ | *Paracoccus*<br><br>*Desulphovibrio* |
| *Nitrogen fixers* | | |
| Enables organisms to grow in nitrate-poor environments, but the energy demand is high | $N_2 + 3FdH_2 \rightarrow 2NH_3 + 3Fd$<br>(*$FdH_2$/Fd: reduced and oxidised ferredoxin) | *Azotobacter*, and many others |

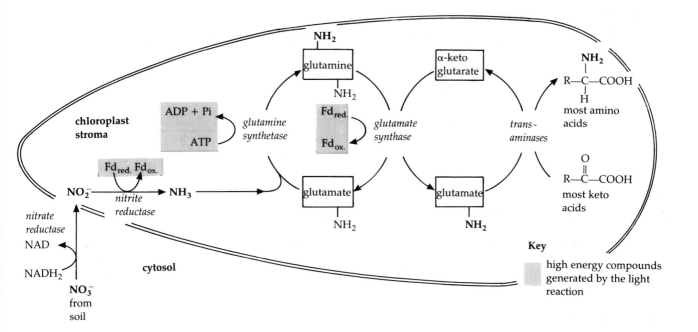

Fig. 5.14 *Nitrogen assimilation in chloroplasts*. Only one of several possible pathways is shown. In some cases, $NADPH_2$ can replace ferredoxin, and nitrogen assimilation can undoubtedly occur outside chloroplasts (as is obviously the case in fungi such as *Neurospora*). Whilst it has long been held that $NH_3$ is directly incorporated into α-ketoglutarate by *glutamate dehydrogenase* to form glutamate, calculations suggest that the level of $NH_3$ needed to favour synthesis by this method would be toxic. *Glutamate dehydrogenase* is therefore much more likely to be involved in deamination during the entry of amino acids into the Krebs cycle.

In photosynthetic cells, nitrite reduction is limited to the chloroplasts and requires **reduced ferredoxin** ($Fd_{red}$). This is a redox agent produced by the light reaction. In these cells, therefore, the reduction of $NO_2^-$ to $NH_4^+$ depends on light as much as photosynthesis itself. $NH_4^+$ is maintained at low, non-toxic levels because it is rapidly incorporated into carboxylic acids to form amino acids. $^{15}NH_4^+$ labelling experiments demonstrate that quantitatively the most important pathway is via incorporation into **glutamate**. This then transfers the amino group to a variety of other carboxylic acids by specific **transaminases**, and so produces corresponding amino acids (Fig. 5.14).

Animals completely lack $NO_3^-$ and $NO_2^-$ reduction systems, and the toxicity of $NH_4^+$ (Fig. 5.14, caption) effectively means that their nitrogen source must be in the 'ready-made' non-toxic form of ingested protein. Their somewhat limited ability to transaminate further requires that some amino acids (the essential amino acids, e.g. methionine) must be specifically present in the diet.

## 5.4.2 Sulphur

Sulphate, like nitrate, is absorbed into root cells by active transport. Some may be assimilated into organic material in the roots themselves, but a substantial quantity appear to be translocated directly to the leaves via the xylem. Assimilation involves three steps (Fig. 5.15):

(i)   activation by ATP;
(ii)  reduction by **sulphate** and **sulphite reductase** systems;
(iii) transfer of the −SH group which has been formed onto serine, forming cysteine.

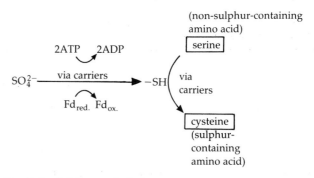

Fig. 5.15 *Sulphate assimilation*

From the latter it can enter into the production of other sulphur-containing substances, such as methionine, biotin, thiamin and CoA. Animals have no capacity whatever for $SO_4^{2-}$ utilisation, and depend entirely upon appropriate sulphur-containing amino acids in their diet.

## 5.5 PHOTORESPIRATION

Two rather curious observations in the 1950s and 1960s revealed a completely unexpected biological process:

(i) If a C3 plant photosynthesising under conditions corresponding to a bright summer day is suddenly put in the dark, it does not immediately settle down to a rate of $CO_2$ evolution corresponding to respiration (Fig. 5.16(iii)). Instead there is a higher-than-expected initial burst of $CO_2$ evolution (Fig. 5.16 (ii)) which persists for several seconds, or even for a few minutes.

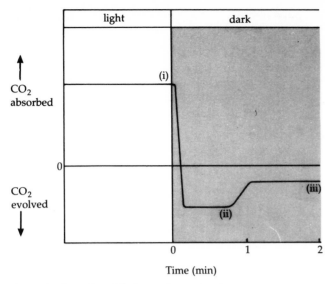

Fig. 5.16 *Transient $CO_2$ burst observed in C3 plants when placed in the dark (after Decker). In these experiments $CO_2$ is normally measured using an infrared analyser. $CO_2$ absorbs infrared, and when a beam is transmitted through air the amount of infrared absorbed is proportional to the amount of $CO_2$ present.*

---

**Q32** Why does Fig. 5.16 show negative $CO_2$ evolution in region (i)?

---

(ii) If a C3 plant in bright light is subjected to a gas flow containing zero $CO_2$ then, not surprisingly, photosynthesis is inhibited. $CO_2$ may be detected in the effluent from the plant, but this is higher than the observed rate of respiration as measured in the dark (Fig. 5.17).

Similar experiments using $^{18}O_2$ show that light stimulates oxygen uptake as well as carbon dioxide evolution. This phenomenon, which is confined to photosynthetic cells, is called **photorespiration**. However, it has nothing whatever to do with true respiration, and it is entirely coincidental that the gases exchanged are the same. Neither has it much to do with photosynthesis, although its origin can be traced back to the key enzyme of the Calvin cycle, *ribulose bisphosphate carboxylase*.

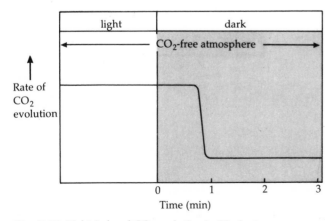

Fig. 5.17 *Light-induced $CO_2$ evolution in C3 plants*

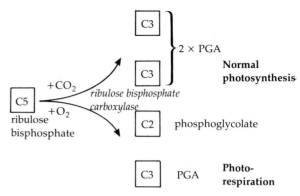

Fig. 5.18 Catalytic activity of RbP carboxylase

It can be shown that oxygen can replace carbon dioxide as a substrate for this enzyme, i.e. $O_2$ is a competitive inhibitor with respect to $CO_2$ (Chapter 2.3.5). As the oxygen level rises above 5% to atmospheric levels (21%) and beyond, the active site on the enzyme becomes progressively utilised for RbP *oxidation* rather than *carboxylation* (Fig. 5.18).

RbP oxidation yields a C3 fragment (phosphoglyceric acid, PGA) and a C2 fragment (**phosphoglycolate**). The former is a normal product of the Calvin cycle. However, extensive research has failed to demonstrate any functional role for glycolate, and it is generally regarded as biologically useless. What amounts to a biochemical salvage operation now occurs to recover as much organic carbon from it as possible. As Fig. 5.19 shows, the recovery process is extremely complex. Fortunately we can ignore nearly all the detail in Fig. 5.19 and still appreciate what is happening (Q33).

---

**Q33** By ignoring the organelles involved, and by assuming that all roads from glycolate lead back to PGA, we can summarise photosynthesis and photorespiration as follows:

**Normal photosynthesis:**
$$2 \times C5 + 2CO_2 \longrightarrow 4 \times C3 \text{ (PGA)}$$

**Photorespiration:**
$$2 \times C5 + 2O_2 \longrightarrow 2 \times C3 \text{ (PGA)}$$
$$+$$
$$2 \times C2 \text{ (phosphoglycolate)}$$

$\hookrightarrow CO_2$

NADH$_2$  NAD

$\hookrightarrow$(via $\longrightarrow$ C3
intermediates)  (PGA)

(i) Why might it be unreasonable to assume that all the glycolate finds its way to PGA?
(ii) Given two RbPs, how many PGAs are ultimately produced as a result of (a) photosynthesis, (b) photorespiration?
(iii) How many organic carbon atoms are lost as $CO_2$, as a percentage of the initial RbP?
(iv) What is the nett loss during photorespiration in terms of energy?

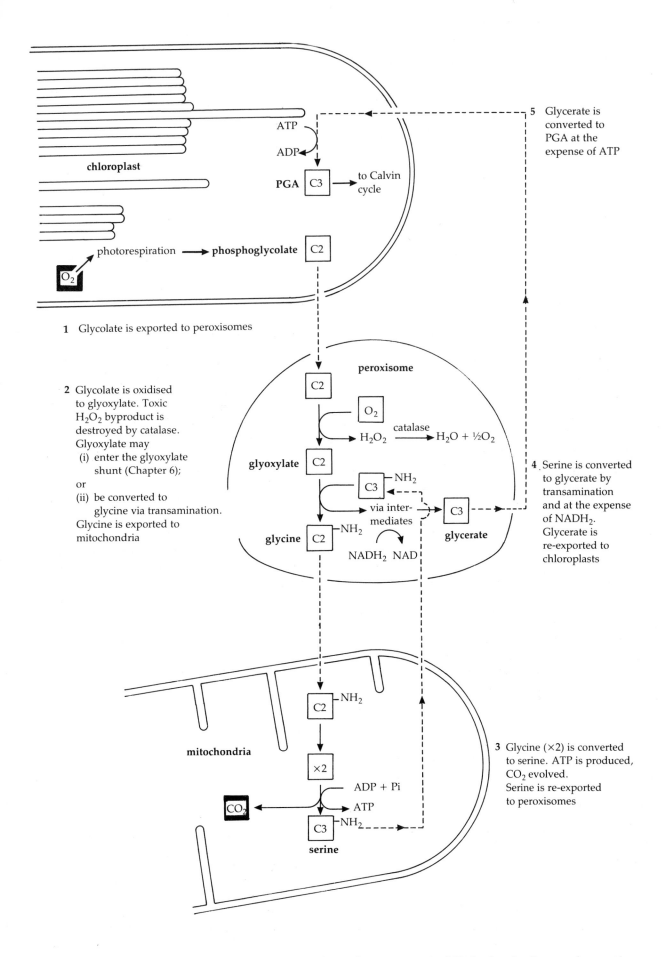

ATP
ADP

**chloroplast**

**PGA** $\boxed{C3}$ → to Calvin cycle

photorespiration → **phosphoglycolate** $\boxed{C2}$

$\boxed{O_2}$

**5** Glycerate is converted to PGA at the expense of ATP

**1** Glycolate is exported to peroxisomes

**2** Glycolate is oxidised to glyoxylate. Toxic $H_2O_2$ byproduct is destroyed by catalase. Glyoxylate may
  (i) enter the glyoxylate shunt (Chapter 6); or
  (ii) be converted to glycine via transamination. Glycine is exported to mitochondria

**peroxisome**

$\boxed{C2}$

$\boxed{O_2}$

$H_2O_2$ — catalase → $H_2O + \frac{1}{2}O_2$

**glyoxylate** $\boxed{C2}$

$\boxed{C3}$ — $NH_2$

via inter-mediates

$\boxed{C3}$ **glycerate**

**glycine** $\boxed{C2}$ — $NH_2$

$NADH_2$  $NAD$

**4** Serine is converted to glycerate by transamination and at the expense of $NADH_2$. Glycerate is re-exported to chloroplasts

$\boxed{C2}$ — $NH_2$

**mitochondria**

$\boxed{\times 2}$

$ADP + Pi$
$ATP$

$\boxed{CO_2}$ ←

$\boxed{C3}$ — $NH_2$

**serine**

**3** Glycine ($\times 2$) is converted to serine. ATP is produced, $CO_2$ evolved. Serine is re-exported to peroxisomes

Fig. 5.19 *The metabolism of glycolate.* A summary of this complex pathway is given in Q33. So that the diagram does not become even more complicated than it already is, the glyoxylate shunt (**2**(i)) has been omitted.

As Q33 suggests, photorespiration is entirely wasteful:
(i) $CO_2$ is robbed of its position on the active site of *RbP carboxylase* and the rate of fixation is reduced.
(ii) Each time $O_2$ replaces $CO_2$ and oxidises RbP, 10% of the existing organic carbon in the latter is lost.
(iii) The light reaction is robbed of reduced dinucleotides, and therefore energy is wasted.
Labelling experiments indicate that in warm sunny conditions photorespiration reduces the nett gains due to photosynthesis by up to 50% in C3 plants, and there is a corresponding reduction in crop yield.

---

**Q34** Formulate hypotheses to explain why photorespiration should be worse when it is (i) very sunny, (ii) warm.

---

Why has natural selection not produced a more efficient *RbP carboxylase*, i.e. one which cannot accept $O_2$ as an alternative to $CO_2$? Perhaps *RbP carboxylase* is as efficient as it can be. Certainly plants living in high risk environments, such as the tropics, have got round the problem by using a better initial fixer of $CO_2$: *PEP carboxylase* (Sections 5.3.2 and 5.3.3).

C4 and CAM plants are so efficient at grabbing $CO_2$ from the atmosphere at low concentrations and creating locally high $CO_2$ levels for *RbP carboxylase*, that they exhibit no detectable symptoms of photorespiration. The identification of glycolate in C4 and CAM plants suggests that photorespiration may in fact occur, but that its effects are largely neutralised by *PEP carboxylase*.

*Photorespiration and agriculture*

The growth rate of C4 plants is considerably higher than that of C3 plants in tropical conditions. This is particularly true when they are compared with agricultural and horticultural C3 plants imported into the tropics from Europe. Not only do the local C4 equivalents exhibit no apparent photorespiration, but they make better use of available water and have a higher optimum temperature (Table 5.3).

The C4 character is a complex one controlled by a large number of genes, and attempts by plant breeders to incorporate C4 characteristics into agriculturally important C3 plants have met with little success. Such breeding programmes may not even be desirable, because the protein content of the C3 leaf is often two to three times greater than that of the C4 leaf. This means two things. On the one hand, C4 plants make more efficient use of nitrogen ($NO_3^-$ deficiency is common in semi-arid regions). On the other hand, animals (and people) are likely to get more protein per mouthful from a C3 plant than from a C4 plant. Since protein deficiency is often a problem in the tropics, there may be advantages in growing the former.

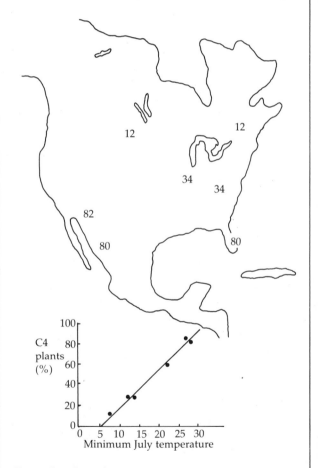

# Highways and Byways of Metabolism

**SUMMARY**

The interconversion of many metabolites is not only possible but essential. Polysaccharide, lipid and amino acid synthesis and degradation are described. The oxygen debt and the special problem of converting lipids to carbohydrates are discussed.

The $CO_2$ fixed into simple organic acids and sugars during photosynthesis rapidly finds its way into a variety of carbohydrates, fats, proteins and nucleotides. To a greater or lesser extent, these four main categories of substances are interconvertible (Fig. 6.1). In this closing chapter we shall look at some of the major syntheses and interconversions.

## 6.1 CARBOHYDRATES

Hexoses such as glucose and fructose are highly reactive and soluble molecules. These features are exploited by organisms, which use them as substrates for respiration, as a basis for the construction of other molecules, and, in animals, as the form in which carbohydrate is transported around the body. However, these same features make hexoses potentially very dangerous and capable of disrupting the delicately balanced biochemical and osmotic properties of an organism. Their intracellular and extracellular concentrations must therefore be carefully controlled. Several alternatives exist for preventing either an undesirable accumulation or a deficiency. The reader may already be familiar with at least one mechanism (Q1).

**Blood glucose levels.** In humans the blood glucose level varies between 60–100 mg per 100 cm$^3$ of blood, but may rise by 50% after a meal. It is brought back to within normal limits by the action of a hormone.

**Q1** Name the organ responsible for detecting the blood glucose level, and the hormone which this organ produces.

This hormone acts on a variety of tissues, especially the liver (L), muscle (M) and adipose tissue (A). Its overall effect is to lower the blood glucose level as a result of:
(i) increased glucose uptake from the plasma into cells (A,M);
(ii) increased conversion of glucose to
    (a) glycogen (L,M)
    (b) fat (A,L)
    (c) proteins (L,M);
(iii) decreased glucose synthesis from organic acids (gluconeogenesis, Section 6.1.3)(L).

**Q2** What role does **glucagon** play in regulating blood glucose levels?

Several other hormones, such as adrenalin and vasopressin (ADH), also stimulate a rise in blood glucose levels. Insulin is the only hormone capable of lowering it.

We shall look first at polysaccharide synthesis and degradation. Since polysaccharides (Fig. 6.2) are less reactive and virtually insoluble, their synthesis is undoubtedly one of the most important ways of effectively removing excess hexoses from metabolism. Moreover, polysaccharides have important properties which make them particularly suitable as food reserves (starch, glycogen, inulin) or skeletal materials (cellulose, chitin).

**Q3** List some properties of starch and glycogen which make them suitable as food reserves.

**Q4** What is chitin? Where is it found? What properties make cellulose and chitin ideal skeletal materials?

**Q5** What other functions do polysaccharides (and oligosaccharides) have in organisms, besides acting as food reserves or skeletal materials?

### 6.1.1 Polysaccharide synthesis

There are three essential features with respect to the synthesis of all the main polysaccharides:
(i) *The synthetic reactions are all intrinsically endergonic* and must be forced along at the expense of nucleotide triphosphates (ATP, UTP or GTP).
(ii) *Nucleotide diphosphates (ADP, UDP, GDP) act as 'energisers'* for the glucose which is being consumed for synthesis.
(iii) *The 'energised' glucose is enzymatically transferred to a pre-existing polysaccharide 'starter'.* The enzymes responsible vary, depending upon the polysaccharide and complex involved, but they are collectively called *glycosyl transferases*. They are found in the cytosol (starch, glycogen), chloroplasts (starch) and outside the plasma membrane (cellulose).

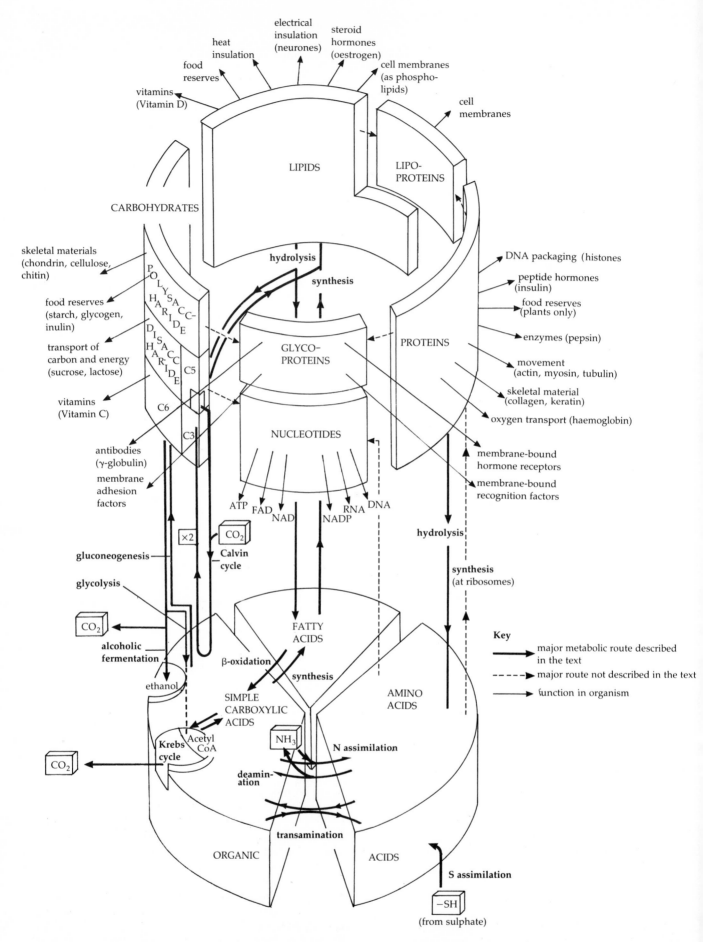

Fig. 6.1 *Major metabolic routes.* A single arrow does not necessarily mean that there is only one pathway: for example, there are about three pathways for the breakdown of fats. Neither does the existence of an arrow mean that a pathway will operate in all organisms or at all times. A nett conversion of fats to carbohydrates, for example, can occur in plants but not in animals.

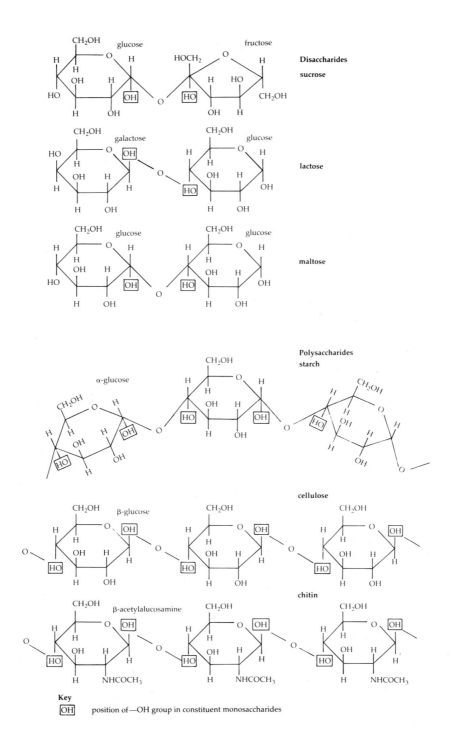

Key

[OH]    position of—OH group in constituent monosaccharides

Fig. 6.2 *Some common disaccharides and polysaccharides*

Readers requiring more specific details are referred to Fig. 6.3 (top diagram). Synthesis of the disaccharide sucrose and that of a 'lesser polysaccharide', inulin, also occurs by similar mechanisms (Fig. 6.3, middle, lower diagrams).

## 6.1.2 Polysaccharide degradation

The routes by which starch and glycogen reserves are degraded are shown in Fig. 6.4. Routes 1 and 2 are the main pathways for the intracellular utilisation of glycogen and starch. The main features are as follows.

(i) *Degradation is intrinsically exergonic* (cf. synthesis), so glucose phosphates are produced by the consumption of inorganic phosphate, not ATP. Hence the enzymes are *phosphorylases* (not *kinases*).

(ii) *The phosphorylases may be subject to allosteric control*: high cAMP levels in the cytosol indirectly convert relatively inactive *phosphorylase b* into more active *phosphorylase a* via the enzyme cascade previously described (Chapter 3.2.3).

Route 3 rarely if ever occurs in cells, but is the principal mechanism for the digestion of dietary starch and starch reserves in seeds.

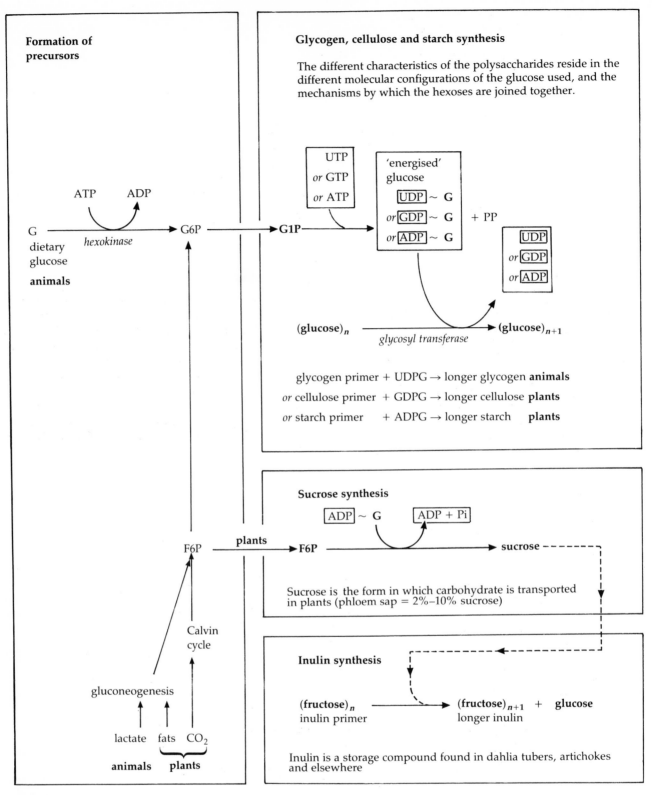

**Key**

| UTP, GTP, ATP, | uridine, cytosine, and adenosine |
| UDP, GDP, ADP | triphosphates and diphosphates |

| G, G1P, G6P | glucose, glucose-1-phosphate, glucose-6-phosphate |
| F6P | fructose-6-phosphate |

Fig. 6.3 *The synthesis of some common disaccharides and polysaccharides*. Since in all cases the hydrolysis reaction (polysaccharide → hexoses) is exergonic, it follows that synthesis can never occur by a simple condensation of hexoses. Glucose + fructose → sucrose, for example, is energetically impossible. In all cases 'energised' hexoses such as ADPG must be employed. Inulin synthesis is an interesting and unusual case, since sucrose can be thought of as an 'energised' form of fructose (Table 1.2). Cellulose synthesis may use UDPG as an alternative to GDPG.

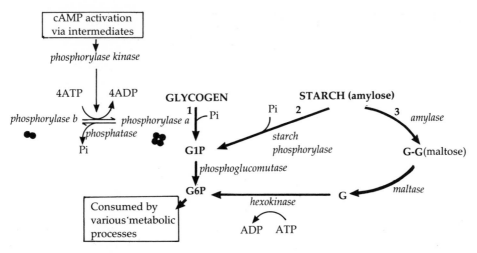

Fig. 6.4 *Degradation of starch and glycogen*

### 6.1.3 Gluconeogenesis

During heavy exercise the demand for ATP outstrips the rate at which it can be produced by oxidative respiration. In order to bridge the 'energy gap', vast quantities of glucose are broken down to lactate. The quantity has to be vast, because glycolysis only yields two ATPs per glucose. As a result, glucose levels in the body may become abnormally low, whilst at the same time the lactate level may become dangerously high. The body therefore needs a system to raise the former and lower the latter. Such a system not only exists, but involves a most remarkable and efficient piece of collaboration between five distinct tissues: muscle, liver, blood, pancreas and adipose tissue. Lactate is first transported by the blood from skeletal muscle to the liver. Here 80% or more is converted back into glucose by a mechanism called **gluconeogenesis**. The rest is respired by the Krebs cycle. Newly formed glucose is used to restore the glucose level of the blood, and to replenish the glycogen stores in liver and muscle. The whole cycle is called the **Cori cycle** (Fig. 6.5).

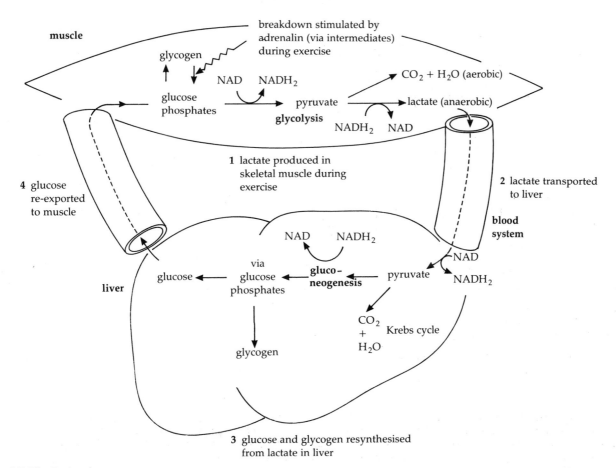

Fig. 6.5 *The Cori cycle*

For gluconeogenesis to occur, three major obstacles must be overcome.

### (i) Irreversible reactions

Some of the steps in glycolysis are irreversible. In the liver, two irreversible steps are of consequence during gluconeogenesis:

$$F6P \xrightarrow[PFK]{ATP \quad ADP} F1:6bP$$

$$PEP \xrightarrow[pyruvate\ kinase]{ADP \quad ATP} pyruvate$$

(There is actually a third irreversible step in glycolysis:

$$G \xrightarrow[glucokinase]{ATP \quad ADP} G6P$$

but the liver enzyme, *glucokinase*, is virtually inactive at low glucose levels and is therefore of little consequence in the glucose-depleted body which exists after exercise.)

In order for gluconeogenesis to occur, these irreversible steps must be bypassed (Q6).

---

**Q6** Look back to Fig. 3.11 and the accompanying text (Chapter 3.5). How are these irreversible steps bypassed? How is it energetically possible?

---

### (ii) Enzyme regulation

If all the glycolytic *and* gluconeogenic enzymes are operating at the same time, then the metabolites in the pathway will be broken down as quickly as they are made, resulting in **futile cycles**. In order to achieve a nett synthesis of glucose, *PFK* and *pyruvate kinase* must be inhibited, and the gluconeogenic enzymes must be activated. This is accomplished through the action of the hormone glucagon. Low blood sugar levels stimulate the pancreas to secrete glucagon, which acts on the liver causing it to produce cAMP. By way of intermediate reactions the latter then:
(a) inhibits both *PFK* and *pyruvate kinase*, slowing glycolysis;
(b) activates the *phosphatases*, stimulating gluconeogenesis;
(c) breaks down residual glycogen in the liver to glucose.
As a result, lactate is converted to glucose ((a) + (b)), and glucose formed from glycogen is not degraded further ((a) + (c)). Gradually the body returns to normal.

### (iii) Energy

None of the above is possible without energy. Since glycolysis produces ATP, it is hardly surprising that gluconeogenesis requires it (Q7).

---

**Q7** Study Figs. 3.11 and 4.1. How many ATPs are required to synthesise one glucose. At what point(s) are they required? (Assume for simplicity that GTP ≡ ATP.)

---

This energy requirement is met by oxidative respiration using the Krebs cycle. The latter is 'fed' not by glucose (which is in short supply), nor by lactate (which is mostly converted to glucose), but by fatty acids which have been mobilised by the action of glucagon on adipose tissue and liver.

### The oxygen debt

For the reaction glucose → 2 lactate, $\Delta G^{o'} = -200$ kJ mole$^{-1}$ (Chapter 4). It follows from this that an energy input is required for gluconeogenesis.

The term **oxygen debt** is often used to describe the amount of oxygen needed to convert lactic acid back into glucose, but we must be careful not to misunderstand this term. Although the first step is indeed an oxidation:

$$\underset{lactate}{C_3H_6O_3} \xrightarrow{NAD \quad NADH_2} \underset{pyruvate}{C_3H_4O_3}$$

the NADH$_2$ which is formed is subsequently used during gluconeogenesis to reduce acid intermediates (—COOH) to sugars (aldehydes, —CHO).

Indeed, *during gluconeogenesis there is no nett oxidation of lactate at all*:

$$\underset{lactate}{2C_3H_6O_3} \xrightarrow[6ATP \qquad 6ADP + 6Pi]{} \underset{glucose}{C_6H_{12}O_6}$$

However oxygen *is* required, in order to produce ATP by oxidative phosphorylation. Hence the oxygen debt is more precisely described as the amount of oxygen needed during the oxidation of fatty acids to provide ATP for the conversion of lactate to glucose.

## 6.2 LIPIDS

Whilst carbohydrates (starch, inulin) are the major food reserves of plants, this is not true of animals. Liver, by far the biggest carbohydrate reserve in mammals, contains less than 6% glycogen. Under fasting conditions a human adult could survive for about one day on the carbohydrate store. In contrast, the cells of adipose (fatty) tissue contain up to 99% fat by weight, providing about a six-week energy reserve.

Lipids are rather variable molecules (Fig. 6.6). Their value as an energy reserve arises from the fact that they are insoluble (hence biochemically and osmotically inert) and much more reduced than carbohydrates. Because they are so reduced, much more oxygen is used during their respiration than for the respiration of carbohydrates. Consequently, more energy is released for ATP synthesis. Thus for a fatty acid:

$$\underset{(palmitic\ acid)}{C_{16}H_{32}O_2} + 23O_2 \longrightarrow 16CO_2 + 16H_2O$$

($\Delta G^{o'} = -9870$ kJ mole$^{-1}$)

whilst for 16 carbohydrate-carbons:

$$2.66C_6H_{12}O_6 + 16O_2 \longrightarrow 16CO_2 + 16H_2O$$

($\Delta G^{o'} = -7650$ kJ mole$^{-1}$)

## Fatty acids

A fatty acid is composed of a carboxyl group (—COOH) bound to a hydrocarbon chain. Most fatty acids contain between 14 and 18 carbon atoms. If the carbons in the chain are joined by single bonds (below) the molecule is described as **saturated**. It will be straight. If a double bond is present then the fatty acid is described as **unsaturated**. It will be kinked.

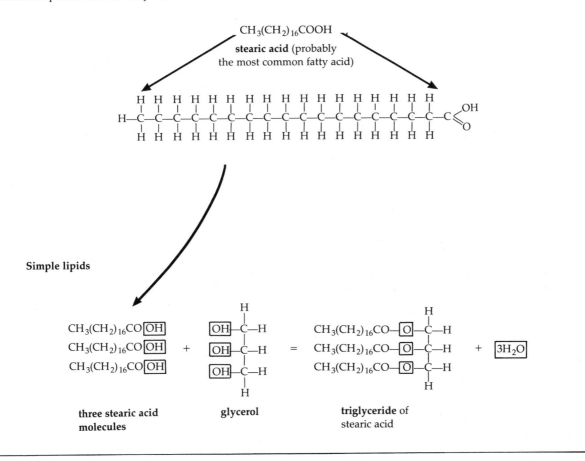

$$CH_3(CH_2)_{16}COOH$$

**stearic acid** (probably the most common fatty acid)

### Simple lipids

$CH_3(CH_2)_{16}CO\boxed{OH}$
$CH_3(CH_2)_{16}CO\boxed{OH}$   +
$CH_3(CH_2)_{16}CO\boxed{OH}$

**three stearic acid molecules**

$\boxed{OH}$—C—H
$\boxed{OH}$—C—H      =
$\boxed{OH}$—C—H

**glycerol**

$CH_3(CH_2)_{16}CO$—$\boxed{O}$—C—H
$CH_3(CH_2)_{16}CO$—$\boxed{O}$—C—H   +   $\boxed{3H_2O}$
$CH_3(CH_2)_{16}CO$—$\boxed{O}$—C—H

**triglyceride** of stearic acid

## Complex lipids

Triglycerides (simple lipids) are often modified by the substitution of phosphate for one or more fatty acids, the replacement of glycerol by sphingosine, and the addition of groups such as choline. Many other modifications also occur.

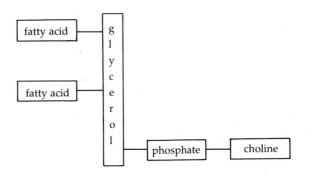

**A phospholipid**
Phospholipids are the most common constituents of cell membranes.

Fig. 6.6 *The structure of lipids.* Lipids play a variety of roles in the body other than as energy reserves (Fig. 6.1). These varied functions account for the enormous diversity in lipid structure.

Information concerning the nature of the respiratory substrate which an organism is using can be obtained by calculating its **respiratory quotient** (Review Question 1).

## 6.2.1 Lipid synthesis

Personal experience provides often too ample evidence that sugars and other carbohydrates can be converted to fats. The first step is the formation of fatty acids from acetyl CoA. It is a rather complex operation involving a large multi-enzyme complex called *fatty acid synthetase*, located on the membrane of the SER (especially of liver cells). In plants, fatty acid synthesis occurs mainly in the chloroplast stroma. The process is summarised in the box below for those readers who wish to know more specific details.

Free fatty acids rarely accumulate in the cell, and are normally incorporated into fats or phospholipids. This takes place in the SER. Hydrolysis of CoA which is attached to the newly synthesised fatty acids releases energy to drive the reaction:

$$3 \; \boxed{C_n} \text{–CoA} \;+\; \text{glycerol} \xrightarrow[\text{intermediates}]{\text{via}} \text{fat} + 3\text{CoA}$$

newly synthesised fatty acid / phosphate from GAP in glycolysis / triglyceride ester ('simple lipid')

## 6.2.2 Fat degradation

There is little free fat in the cytosol. It is normally restricted to membrane-lined vesicles in the cytoplasm. In animals a small amount of dietary fat also circulates

---

### Fatty acid synthesis

The complexity of fatty acid synthesis (Fig. 6.7) may be reduced to four basic steps.

(i)  Conversion of sugars to acetyl CoA.
(ii)  Carboxylation of acetyl CoA to form a 'carbon primer', malonyl CoA

$$\boxed{C2} + CO_2 \xrightarrow[\substack{\text{vitamin, biotin} \\ ATP \;\; ADP + Pi}]{\text{via the } CO_2\text{-binding}} \boxed{C3}$$

acetyl COA / malonyl CoA

ATP is supplied by the light reaction of photosynthesis, or by respiration.

(iii)  Condensation of acetyl CoA(C2) with malonyl CoA(C3) to yield a C4 fatty acid:

$$\boxed{C2} + \boxed{C3} \rightarrow \boxed{C4} + CO_2 + CoA$$

acetyl CoA / malonyl CoA / C4–CoA

(iv)  Reduction of the C4 fragment, which subsequently acts as a 'primer' for the addition of more C2 fragments from malonyl CoA:

$$\boxed{C4} + \boxed{C3} \xrightarrow{2NADPH_2 \;\; 2NADP} \boxed{C6} + CO_2 + CoA$$

primer –CoA / malonyl CoA / new primer –CoA

$NADPH_2$ is generated either by photosynthesis or by the pentose phosphate pathway (Chapter 4.1.1, panel).
Further condensation of malonyl CoA with the primer yields even-numbered saturated fatty acids of increasing length:

$$\boxed{C_n} + \boxed{C3} \rightarrow \boxed{C_{n+2}} + CO_2 + CoA$$

fatty acid –CoA / malonyl –CoA / longer fatty acid –CoA

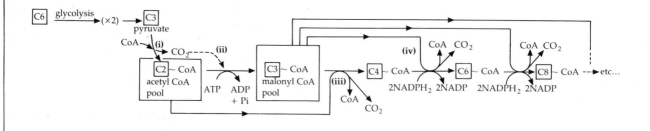

Fig. 6.7 *Fatty acid synthesis*. Each arrow may summarise several intermediate steps.

---

Odd-numbered fatty acids may be synthesised by simply substituting a C4 fragment for malonyl CoA at stage (iii). Unsaturated fatty acids can be obtained by reduction reactions involving $NADPH_2$. The ability of animals to manufacture unsaturated fatty acids is limited, and at least some are required in the diet. These are manufactured ultimately by plants, and are called the **essential fatty acids**.

in the bloodstream. The first step in fat utilisation is its hydrolysis by *lipases*. These are associated with the fat-containing vesicles and the plasma membrane.

$$3H_2O + \text{simple} \xrightarrow{\text{lipase}} \text{glycerol} + \text{3 fatty}$$
$$\text{lipid} \qquad\qquad\qquad\qquad \text{acids}$$

Glycerol is easily dispensed with (Q12).

---

**Q8** Glycerol is phosphorylated and oxidised to GAP (from whence it came (see above)):

glycerol

ATP ⌐ NAD
ADP ↙ ↘ NADH$_2$

via intermediates

glyceraldehyde-3-phosphate (GAP)

Calculate the nett gain in ATP when one molecule of glycerol is oxidised to $CO_2$ and $H_2O$ via glycolysis and the Krebs cycle (Figs. 4.1 and 4.6).

---

*β-oxidation*

Three main metabolic pathways exist for the breakdown of fatty acids, plus several minor ones. The most important pathway, which occurs in mitochondria of plants and animals, is called β-oxidation. The process essentially involves:
(i) transporting the fatty acid from the cytosol through the relatively impermeable mitochondrial inner membrane using a carrier molecule;
(ii) activation of the fatty acid by ATP and CoA:

$$\text{etc.} -C_\gamma-C_\beta-C_\alpha-COOH + CoA$$

ATP
AMP + PP

$$\text{etc.} -C_\gamma-C_\beta-C_\alpha-CO.CoA + H_2O$$

(α, β, γ = first three carbons in the hydrocarbon chain)

(iii) intermediate reactions culminating in a second molecule of CoA cracking off a two-carbon fragment (i.e. between the α and β carbons)

$$\text{etc.} -C_\gamma-C_\beta-C_\alpha-CO.CoA + CoA$$

$C_x$

via intermediates

FAD/NAD
FADH$_2$/NADH$_2$

$$\text{etc.} -C_\gamma-C_\beta O.CoA \quad + \quad -C_\alpha-CO.CoA$$

$C_{x-2}$ — acetyl CoA (C2)

Step (iii) is repeated until the entire fatty acid has been converted into chunks of acetyl CoA. The latter then enters the Krebs cycle. Since the chain becomes progressively shorter and FADH$_2$ and NADH$_2$ are produced (step iii), the pathway is sometimes called the **fatty acid oxidation spiral**.

---

**Q9** During the respiration of one molecule of fatty acid, what 'energy inputs' are required?

**Q10** Assume one molecule of fatty acid containing sixteen carbons (e.g. palmitic acid, $C_{16}H_{32}O_2$) is degraded by β-oxidation. How many (i) acetyl CoAs, (ii) NADH$_2$s and (iii) FADH$_2$s will be produced?

**Q11** In the Krebs cycle, how many ATPs are produced per acetyl CoA consumed?

**Q12** Taking your answers to Q9–Q11 together, how many ATPs are produced per palmitic acid molecule?

**Q13** Given the equation in Section 6.2 for the oxidation of palmitic acid, and assuming one mole of ATP ≡ 30 kJ of energy, what percentage of the energy in palmitic acid is captured as ATP during its complete oxidation? Show your working.

---

### 6.2.3 Fats to carbohydrates

By chemical composition, animals only contain between 1%–5% carbohydrate. This level can normally be maintained by carbohydrate intake from the diet, and there is rarely any need for a nett conversion of fatty food reserves to carbohydrate. By contrast, plant cells are about 20% carbohydrate (f.wt.), largely on account of their cellulose cell walls, and consequently there may be times when a fat-to-carbohydrate conversion is essential. Prior to the onset of photosynthesis, for example, a germinating castor oil seed must mobilise its fatty food reserves and use a significant proportion for the synthesis of cellulose.

Whilst the conversion of carbohydrate to fat presents no unusual biochemical problems, the reverse process presents two problems (see Fig. 6.8 and box below it). Plants and many micro-organisms possess an enzyme system which neatly bypasses both problems. This makes a nett conversion of fat to carbohydrate possible (Fig. 6.9). It works as follows:
(i) Acetyl Co (C2) combines with oxaloacetate (C4) to form citrate (C6), as in the Krebs cycle. This is converted to the isomer isocitrate (C6) which is broken down into C4 and C2 fragments (**succinate** and **glyoxylate** respectively). The latter combines with a *second* acetyl CoA to form a *second* C4 molecule (malate). Hence by this route, the **glyoxylate shunt**, C4 acids are generated without any loss of organic carbon as $CO_2$. This is, of course, in complete contrast to the Krebs cycle. The shunt operates in special membrane-bound cytoplasmic organelles called **glyoxysomes**.

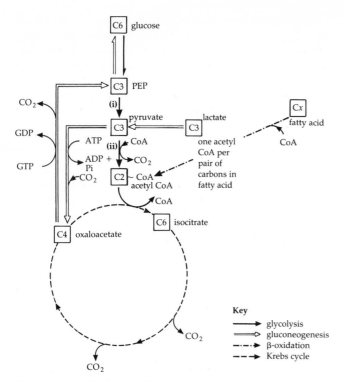

Fig. 6.8 *Carbohydrate and fat metabolism in animals*

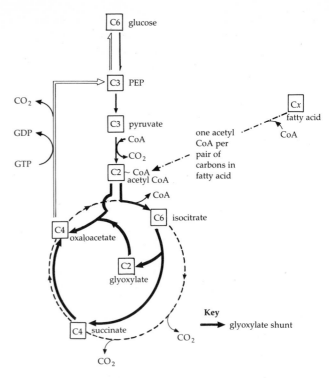

Fig. 6.9 *The conversion of fats to carbohydrates: a pathway found only in plants and micro-organisms*

---

Look at Fig. 6.8:

**Q14** To what substance are fatty acids broken down?

*Problem 1*

The C2 fragment in acetyl CoA cannot be converted directly to PEP (and ultimately to carbohydrate) because reactions (i) and (ii) (Fig. 6.8) are irreversible (strongly exergonic).

**Q15** The only route for the further metabolism of acetyl CoA is into the Krebs cycle. How many carbons are lost as $CO_2$ during each turn of this cycle?

*Problem 2*

Hence, in effect, all the carbon entering the Krebs cycle as acetyl CoA is lost as $CO_2$ during a single turn. (It is actually rather more indirect than the diagram suggests.)

---

**glyoxylate shunt**

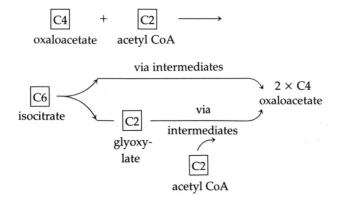

**Krebs cycle**

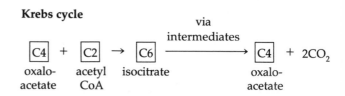

Why don't organisms 'normally' use the glyoxylate shunt and avoid wasting carbon as $CO_2$ altogether? Because the shunt generates no ATP; only the complete Krebs cycle does.

(ii) Oxaloacetate (C4) now provides an escape route from the Krebs cycle, as previously described (Figs. 3.11 and 6.9).

---

**Q16** Animals do possess *PEP carboxykinase* (i.e. can convert oxaloacetate (C4) to PEP (C3)): it is used during gluconeogenesis from lactate. Compare Figs. 6.8 and 6.9 carefully, and suggest why they still cannot make a nett conversion of fatty acids to carbohydrate, even though they possess this enzyme.

---

## 6.3 CARBOXYLIC TO AMINO ACIDS AND *VICE VERSA*

---

**Q17** The assimilation of nitrogen into carboxylic acids to form amino acids was discussed in Chapter 5 (Fig. 5.14). The utilisation of protein and amino acids as substrates for the Krebs cycle was discussed in Fig. 4.7. Collate the main points of both topics into a set of short notes.

Many of the twenty amino acids which occur in proteins are formed by **transamination** reactions involving glutamate. The reactions are reversible, and also serve to convert surplus amino acids into carboxylic acids for the Krebs cycle:

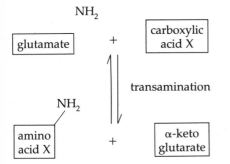

**Transaminases** are specific for particular carboxylic acids and their corresponding amino acids; this allows the process to be finely controlled. Not all amino acids are synthesised via glutamate; some arise by amination of other organic acids. Amination of pyruvate, for example, yields alanine, whereas deamination of the latter regenerates pyruvate. There is some variation between animals and plants in how particular amino acids arise, and even more diversity amongst micro-organisms.

Discussion of amino acid formation raises the question of how they are incorporated into proteins. The major issues involved in this problem are genetic rather than metabolic and enzymic, and readers who are interested should consult *Genetic Mechanisms* in this series.

---

**Study guide**

*Vocabulary*

What are the functions of:

| | |
|---|---|
| β-oxidation | glyoxysomes |
| *glycosyl transferases* | the pentose phosphate pathway |
| gluconeogenesis | *transaminases* |

*Review Question*

1 *The respiratory quotient*
The respiratory quotient is the amount of $CO_2$ given off divided by the amount of $O_2$ absorbed during respiration:

$$RQ = \frac{CO_2 \text{ produced}}{O_2 \text{ absorbed}} \text{ per unit time}$$

(i) What would be the RQ for a tissue that is respiring:
 (a) glucose, by alcoholic fermentation
 (b) glucose, by aerobic respiration
 (c) fats, such as palmitate (Section 6.2)?
 Respiration of proteins gives an RQ similar to fats.
 Values greater than 1 suggest a mixture of aerobic and anaerobic respiration, values just under 1.0 are more difficult to interpret but may indicate that a mixture of fats or proteins is being respired along with carbohydrates.
(ii) Graph the following results which show the changes occurring in the organic constituents of a sample of castor oil seeds during germination.

| Days | 0 | 4 | 6 | 8 | 10 | 15 |
|---|---|---|---|---|---|---|
| Endosperm fat (g) | 383 | 355 | 142 | 57 | 28 | 20 |
| Endosperm carbohydrate (g) | 28 | 50 | 170 | 139 | 2 | 2 |
| Embryo fat (g) | 1.0 | 0.5 | 1.0 | 1.0 | 1.5 | 2.0 |
| Embryo carbohydrate (g) | 2.5 | 6.5 | 14 | 57 | 241 | 497 |
| RQ | – | ≪1 | ≪1 | <1 | ≤1 | 1 |

(iii) What is the percentage change in weight over (a) the first 5 days
 (b) the next 5 days
 (c) the last 5 days
 (Express the values as percentages of the weight at the start of each of the three intervals.) Explain these changes.
(iv) Suggest an explanation for the change in the composition of the endosperm during the first week. Briefly outline the mechanism involved.
(v) Comment on the inverse relationship between the endosperm and embryo carbohydrate levels.
(vi) What is likely to be the major respiratory substrate between days (a) 0–5, (b) 5–10 and (c) 10–15?
 Give reasons for your answers.

2 Why do athletes continue to 'gasp for breath' *after* a race is *over*?
Explain the following graph in terms of the metabolism of carbohydrates and fats.

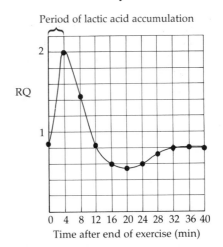

Fig. 6.RQ2 *Changes in RQ during and after exercise.* The RQ for a resting human adult is approximately 0.85. A person undertook violent exercise for 30 seconds and the RQ was measured at regular intervals afterwards. The results are shown on the graph above.

Light is a form of electromagnetic radiation. All such radiation has wave-like properties. The distance between any two similar points in successive waves is called the **wavelength** and is measured in nm. Only electromagnetic radiation between about 450 nm (violet) and 700 nm (red) is visible light (Fig. A.1).

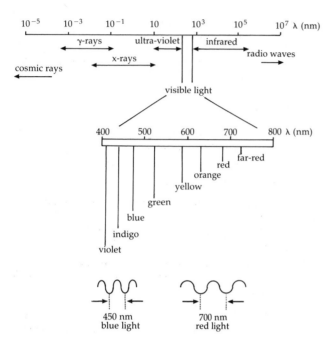

Fig. A.1 *Characteristics of the electromagnetic spectrum*

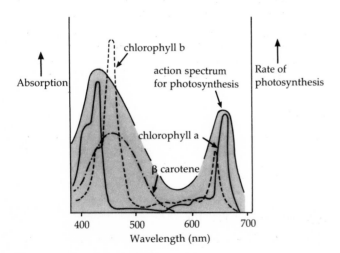

Fig. A.2 *Absorption and action spectra*

Chemicals absorb characteristic wavelengths. Chlorophyll, for example, absorbs mostly blue and red. The wavelengths absorbed by a substance constitute the **absorption spectrum** of that substance. Such spectra are so distinctive that a sample of unknown material can be identified from its absorption spectrum (Fig. A.2). Similarly, if the wavelengths which cause some process to occur are known (the **action spectrum**), then the pigments involved may be inferred.

Another important property of electromagnetic radiation is that it is transferred in discrete units called **quanta** (or **photons** in the case of the visible spectrum). Planck showed that the amount of energy transferred per quantum varied, and depended upon the number of waves transmitted in a given time. Since short wavelength light delivers more waves per second than long wavelength light, it has more energy per quantum (Fig. A.1),

i.e. energy per quantum $(E) \propto \dfrac{1}{\text{wavelength}}$

or more precisely

$$E = h\frac{c}{\lambda} \;(J)$$

where $h$ = Planck's constant ($6.626 \times 10^{-34}$ J s)
  $c$ = speed of light ($3 \times 10^{17}$ nm s$^{-1}$)
  $\lambda$ = wavelength (nm).

The first step in a photochemical reaction occurs when an electron in a molecule becomes energised by absorbing a quantum of light energy. For a molecule of, say, chlorophyll to undergo a photochemical reaction it must absorb *precisely* one quantum. Similarly, for one mole of a substance ($6.023 \times 10^{23}$ molecules) to become excited it must absorb $6.023 \times 10^{23}$ quanta, or one mole of quanta. Einstein concluded that the amount of energy $(E)$ which one mole of a substance absorbs in a photochemical reaction is therefore:

$$E = h\frac{c}{\lambda} \times N \; J\,mole^{-1}$$

where $N$ = Avogadro's number ($6.023 \times 10^{23}$).

---

**Q1** How much energy is contained in one mole of (i) red light at 650 nm? (ii) blue light at 480 nm? Convert your answers to kJ mole$^{-1}$.
(*Section 5.2 (v):* Is there any significance in the fact that the reduction of NADP to NADPH$_2$ by water requires at least 220 kJ mole$^{-1}$, and that the reaction centres of PSI and PSII absorb mainly in the *red* end of the spectrum?)

---

# Answers

IQ = In-text questions; RQ = Review Questions;
EQ = Extension Questions

## CHAPTER 1

**IQ1** The chemical energy stored in glucose would have been converted into kinetic energy (energy of movement) and heat.

**IQ2** $\Delta G^{o\prime}$ has a '+' sign.

**IQ3** The first law: the reactants are the same in both reactions; only the direction of the arrow is different.

**IQ4** Light, chemical.

**IQ5** 1.13 V; $\Delta G^{o\prime} = -2 \times 96.5 \times 1.13 = -218$ kJ mole$^{-1}$.

**IQ6** $\Delta G = -1$ kJ mole$^{-1}$, i.e. the reaction can proceed.

## CHAPTER 2

**IQ1** See table.

**IQ2** (i) S and E form a complex; (ii) complex is transient; (iii) E is restored intact after reaction; (vi) see Q3 (i).

**IQ3** (i) Equimolar implies equal numbers of molecules. Hence *one* S molecule binds to *one* E molecule, i.e. *one* active site *per* E molecule.
(ii) Since 20 $\mu$m E causes a complete change in absorption spectrum of S at 40 $\mu$m, 1E = 2S, i.e. *two* active sites per E molecule.
In short, enzymes have a small, definite number of active sites (rarely more than 4).

**IQ4** (i) This site *may* be the active site of the enzyme, and the analogue is blocking it.
(ii) The substrate analogue and the true substrate behave in a similar way.

**IQ5** pH, temperature, and substances such as cofactors, coenzymes and inhibitors.

**IQ6** Large amounts of substrate and very short times mean that a negligible amount of reactant is consumed/produced, i.e. the reactant concentration remains (almost) constant.

**IQ7** Slight pH shifts reversibly alter ionisation; violent shifts denature the enzyme by breaking the bonds which determine the tertiary structure.

**IQ8** 30 °C = (i) 5 $\mu$g min$^{-1}$;  (iii) 4.65 $\mu$g min$^{-1}$
50 °C = (ii) 16 $\mu$g min$^{-1}$;  (iv) 1 $\mu$g min$^{-1}$.
(iii) compared with (iv) indicates 50 °C causes rapid denaturation, with high initial velocity (ii) rapidly declining. (i) extrapolated over 10 min (5 $\mu$g min$^{-1}$ × 10) suggests negligible denaturation at 30 °C. Optimum for organism may be near 30 °C. Comparison of (iii) with (i) suggests $Q_{10} \simeq 2$.

**IQ9** (a) = (iii); (b) = (ii); (c) = (i). For reasons, see preceding text.

**IQ10** Place solution of inhibited enzyme in dialysis tubing. Stand in pure water. If inhibition is due to NCI, activity of enzyme will be restored as NCI diffuses out of tubing.

**IQ11** (i) When [S] is high; (ii) when [S] is low.

**IQ12** $v = \dfrac{V_{max}[S]}{[S] + K_m}$

**RQ2** (ii) $V_{max} = 0.45$ mg min$^{-1}$; $K_m = 1.82$ mM.

**EQ** (ii) 10 $\mu$moles min$^{-1}$; 3 mM.
(iii) $10^5$ molecules of substrate per molecule enzyme per minute.

## CHAPTER 3

**IQ1** (i) Phosphorylates using ATP; (ii) phosphorylates using inorganic phosphate.

**IQ2** By '*direct action*': $5 \times 10^3$ G6Ps in 5 s; *by cascade*: $10^3 \times 10^4 \times 10^4 \times 10^4 = 10^{15}$ G6Ps in 5 s.

**IQ3** (i) Lipid-derived; oestrogen (or any other sex hormone);
(ii) Protein hormones work via secondary messengers (cAMP or $Ca^{2+}$).

**IQ4** (i) Feedback inhibition (route z) inhibits $e_4$ (and $e_2$; but a supply of B for C is still possible via isozyme $e_1$).
(ii) Feedback inhibition (route y) inhibits $e_3$ (and $e_1$; but a supply of B for D is still possible via isozyme $e_2$).

**IQ5** By using two phosphorylated nucleotides (ATP, GTP) = $2 \times 30$ kJ mole$^{-1}$, $\Delta G'$ PEP synthesis $> \Delta G'$ PEP hydrolysis ($-54$ kJ mole$^{-1}$). 'Oxaloacetate' is little more than a means to an end.

**RQ1** (iii) *Either*: solution contains two isozymes (each inhibited by lysine or threonine), *or* one enzyme has two different binding sites for these inhibitors, which cooperate. These alternative hypotheses can be distinguished by electrophoresis.

**RQ2** Regulation of enzyme *activity*: quick response, but wasteful of materials and energy when enzyme is inactive, therefore use only if enzyme is always likely to be needed to some extent. (Others by corollary.)

## CHAPTER 4

**IQ1** +30 kJ for synthesis: energy cannot be created or destroyed. Hence if a certain quantity is released in one reaction the same amount must be required in the reverse reaction.

**IQ2** 2.

**IQ3** 60 kJ mole$^{-1}$ (2 × 30).

**IQ4** (i) 60/200 × 100% = 30%; (v) 60/210 × 100% = 29%.

**IQ5** Lost as heat: 70%–71%.

**IQ6** 'Standard' conditions do not exist in cells, hence the preceding calculations may only be roughly indicative of the actual situation.

**IQ7** (i) glucose (S) → glucose-6-phosphate (P); (iii) fructose-6-phosphate (S) → fructose 1:6 bisphosphate (P).

**IQ8** 2.

**IQ9** *Kinases* add phosphate to molecules at the expense of ATP.

**IQ10** $\Delta G°$ may be positive, but $\Delta G$ is not because the product (aldehyde) is subsequently consumed (Chapter 1.3.4).

**IQ11** There are two DPGAs formed per glucose, i.e. four ATPs per glucose, but step 1 consumes 2ATPs, hence nett gain of two ATPs per glucose consumed.

**IQ12–IQ13** Check your answers against Chapter 1.3.2.

**IQ14** $\Delta G°' = -143$ kJ mole$^{-1}$; *no* – oxidation by $O_2$ yields more energy; magnitude is proportional to difference in redox potential, i.e. $O_2$ is more oxidising than $NO_3^-$.

**IQ15** $O_2$; it is theoretically possible to make more ATP than when using $NO_3^-$.

**IQ16** Two.

**IQ17** One turn, since two molecules of $CO_2$ ($2 \times C$) are released, and acetyl CoA = C2.

**IQ18** C6 is split into $2 \times C3$ during glycolysis; each C3, in the form of pyruvate, loses one C (as $CO_2$) during acetyl CoA formation.

**IQ19** Two turns per glucose.

**IQ20** See Fig. 4.6.

**IQ21** A relatively high pH means it is less acid, i.e. $H^+$ is being pumped out of the matrix.

**IQ22** Diffusion of $H^+$ through a leaky membrane results in the pH of both sides ultimately reaching the same value.

**IQ23** The ATP synthesis system might be driven backwards, i.e. ATP could be hydrolysed back to ADP + Pi (this is confirmed by experimental observations).

**RQ1** (i) 4 (2 in glycolysis; 2 in Krebs); (ii) 8 ($4 \times 2$ turns); (iii) 2; (iv) 2; (v) $4 + (8 \times 3) + (2 \times 2) + (2 \times 2) = 36$ATPs; (vi) $30 \times 36/2870 \times 100\% = 37.6\%$; $62.4\% =$ heat; (vii) 2; $2 \times 30/2870 \times 100\% = 2\%$.

## CHAPTER 5

**IQ1** (i) Light at appropriate wavelengths; (ii) light at appropriate intensities; (iii) chlorophyll (in chloroplasts); (iv) $CO_2$; (v) $H_2O$; (vi) enzymes/cofactors; (vii) appropriate temperature; (viii) a supply of C5 sugar to trap $CO_2$.

**IQ3** Mice in an enclosed chamber can only survive if a photosynthesising green plant is also present.

**IQ4** Localised concentration of aerobic bacteria suggests maximal $O_2$ evolution (photosynthesis) at the blue and red regions of the spectrum.

**IQ5** Easier to (i) obtain replicates using identical weights of algae; (ii) detect $O_2$ evolution; (other answers possible).

**IQ6** $mm^3$ $g^{-1}$ $hr^{-1}$ (*volume* per *standard amount of plant* in a given *time*).

**IQ7** (i) $H_2O$ used for many purposes other than photosynthesis. (ii) measuring carbohydrate production involves killing samples, more complex analytical procedure (neither may be possible or desirable); (iii) technically difficult to measure light absorbed.

**IQ8** Rate of $O_2$ uptake (or $CO_2$ production) is not affected by light.

**IQ9** From A–B, $O_2$ is absorbed therefore respiration > photosynthesis.

**IQ11** (i) Light; (ii) $CO_2$.

**IQ12** Confirms $CO_2$ was limiting between $x$–$y$ in I.

**IQ13** Increase $CO_2$ level; raise temperature? (Try it and see!)

**IQ14** No.

**IQ15** (i) Thermochemical; (ii) no; at low temperature the slow thermochemical reaction 'catches up' with the preceding light reaction *only if the dark interval is long enough*.

**IQ16** A series of four histograms, similar to A and B should be drawn.

**IQ17** Partial inhibition of enzymes slows thermochemical reaction; darkness allows it to 'catch up'.

**IQ18** A mass spectrometer (detects isotopes by their *atomic mass*). $^{18}O_2$ is *not* radioactive, so a Geiger counter *cannot* be used.

**IQ19** $6CO_2 + 12H_2^{18}O \rightarrow C_6H_{12}O_6 + 6^{18}O_2 + 6H_2O$.

**IQ20** $C^{18}O_2$ should *not* result in $^{18}O_2$ evolution.

**IQ21** The number of quanta needed to produce one $O_2$; $1/0.12 = 8$.

**IQ22** A, B, C and D all equal 4.

**IQ23–IQ25**

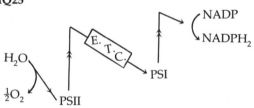

**IQ26** PGA; 3.

**IQ27** $C5 + CO_2 \rightarrow 2 \times C3$; a portion of the C3 pool is re-used to synthesise C5. In the absence of $CO_2$, C5 therefore accumulates and C3 decreases.

**IQ28** Two of each, since 2PGAs are produced per $CO_2$.

**IQ29** One.

**IQ30** 2NADPH$_2$, 3ATP; 2NADPH$_2 \equiv 2 \times 220$ kJ mole$^{-1}$; 3ATP = $3 \times 30$ kJ mole$^{-1}$, so energy required per $CO_2$ = $440 + 90 = 530$ kJ mole$^{-1}$, of which $440/530 \times 100$ (83%) comes from NADPH$_2$, and the remaining 17% from ATP.

**IQ31** For C6, $6 \times 530$ (Q30) = 3180 kJ mole$^{-1}$; efficiency = $2870/3180 = 90\%$.

**IQ32** Photosynthesis > respiration in region (i).

**IQ33** (i) Some may enter the glyoxylate shunt (Chapter 6): Fig. 5.19. (ii) 4; 3.

(iii) One $CO_2$ per 2 RbPs, i.e. 10% loss.

(v) Each $CO_2$ lost $\equiv$ 3ATP + $2NADPH_2$ originally required to fix it via the Calvin cycle. Additionally, there is a nett consumption of one $NADH_2$ during photorespiration. Total energy loss = $(3 \times 30) + (2 \times 220) + 220 = 750$ kJ mole$^{-1}$.

**IQ34** High rates of photosynthesis may reduce the $CO_2$ concentrations in the leaf to low levels, so encouraging photorespiration. Water stress may induce temporary stomatal closure, so worsening the problem.

## CHAPTER 6

**IQ1** Pancreas; insulin (from β-cells in Islets of Langerhans).

**IQ2** Glucagon is secreted by the α-cells in the Islets of Langerhans when the glucose level is at the lower end of the normal range. Its effect is to raise the glucose level (it is antagonistic to insulin) by promoting glycogen breakdown in liver, and the utilisation of fat rather than carbohydrate for respiration.

**IQ3** Relatively inert (because insoluble): do not interfere with metabolism, and exert negligible effect on OP; easily digested to hexoses when required; compact shape: easily stored.

**IQ4** A polymer of β-acetyl glucosamine (glucose in which —$NHCOCH_3$ replaces —OH on C–2). Belongs to a category of modified carbohydrates called mucopolysaccharides (= glucosaminoglycans). Both chitin and cellulose are insoluble, fibrous, strong, inert, indigestible and capable of cross-linking to other molecules.

**IQ5** Adhesion factors in plasma membranes (bind adjacent animal cells together); molecular 'identity cards' which assist lymphocytes in recognition of self/non-self; antibodies; hormone receptors: all these when oligosaccharides/polysaccharides are linked to protein to form glycoproteins.

**IQ6** (i) By uncoupling F6P → F1:6bP from ATP utilisation the reaction can go in the reverse direction. However, a different enzyme is involved, because at the molecular level it is not *exactly* a reversal (no ATP is involved).

(ii) PEP → pyruvate is bypassed via oxaloacetate at the expense of two ATPs (or one ATP and one GTP) per lactate.

**IQ7** Four ATPs from lactate to PEP, plus two ATPs from 3PGA to DPGA. Total = six ATPs per glucose from lactate.

**IQ8** Via glycolysis: 2ATP + $2NADH_2$ (latter = two ATPs each, see Chapter 4 RQ1(iv)) = six ATPs; via Krebs: 15; total = 21 ATPs, minus one (Q8) = 20 ATPs.

**IQ9** A single ATP is needed at step (ii) for each fatty acid molecule.

**IQ10** (i) 8; (ii) 7; (iii) 7.

**IQ11** 12.

**IQ12** $(12 \times 8) + (7 \times 3) + (7 \times 2) - 1 = 130$ ATPs per C16 fatty acid.

**IQ13** $\dfrac{130 \times 30}{9870} \times 100 = 39.5\%$.

**IQ14** Acetyl CoA/Section 6.2.2.

**IQ15** Two.

**IQ16** Acetyl CoA (2) can only go 'clockwise' round the Krebs cycle, which loses $2 \times CO_2$ per turn. Hence there cannot be a nett gain of C4 from acetyl CoA in animals.

**RQ1** (i) (a) ∞; (b) 1.0; (c) 0.695.

(iii) (a) 10% decrease; (b) further 26% decrease; (c) 206% increase.

(iv) Fat converted to carbohydrate via glyoxylate shunt.

(v) Transfer of sugar from endosperm to embryo between days 6–12.

## APPENDIX

**QA.1** (i) 184 kJ mole$^{-1}$; (ii) 249 kJ mole$^{-1}$. The significant point is that a single photosystem activated by *red* light could not generate enough energy to reduce NADP by $H_2O$ since this requires a minimum of 220 kJ mole$^{-1}$.

# Glossary

(Only short working definitions are given. Some terms are given a fuller treatment in the text, see *Index*.)

*aerobic respiration*
Respiration involving a nett oxidation of food, and in which molecular oxygen acts as the terminal hydrogen acceptor.

*activation energy*
The energy which molecules must possess in order to undergo a chemical reaction.

*alcoholic fermentation*
See *glycolysis*.

*allosteric*
Having no structural resemblance;
— *regulator*: a molecule capable of regulating an enzyme, by binding at a site other than the active site.
— *enzyme*: an enzyme which can be activated or inactivated by allosteric regulators.

*anaerobic respiration*
Any form of respiration in which molecular oxygen is not consumed. It may be *non-oxidative* (as in glycolysis), or *oxidative* (as with *denitrifying bacteria*). In the latter case oxidised components such as soil $NO_3^-$ substitute for molecular oxygen.

*catabolite repression*
Where the utilisation of a potential substrate for a metabolic pathway is inhibited by an alternative substrate.

*compensation point*
The set of conditions at which two processes with opposite overall effects just balance each other.

*covalent bond*
Where two atoms are held together by the sharing of one or more pairs of electrons. Covalent bonds are strong: 200–800 kJ mole$^{-1}$ of energy is required to break them, and they are therefore relatively stable compared with other biologically important bonds.

*disulphide bond*
A covalent bond involving two sulphur atoms (—S—S—).

*endergonic*
Energy-consuming reaction.

*entropy*
A measure of disorder. When a chemical is involved in an exergonic reaction, a proportion of the energy in it is converted into forms which are no longer available for doing useful chemical or physical work. This is called the change in entropy of the system.

*equilibrium*
Where two opposing forces just balance each other:
$$A \rightleftharpoons B$$
In a chemical reaction equilibrium does *not* imply equal amounts of A and B.

*exergonic*
Energy-releasing reaction.

*feedback inhibition*
The situation where the product of a reaction (or pathway) inhibits the enzyme(s) so that the rate of reaction is slowed.

*glycolysis*
Splitting of glucose ($C_6H_{12}O_6$). The end product of the pathway is lactic acid ($C_3H_6O_3$). Occurs in animals under *anaerobic conditions*; in plants, ethanol and $CO_2$ are generally produced as a result of the variant *alcoholic fermentation*. In the presence of oxygen, glycolysis is normally interrupted prior to lactate formation, and the intermediates are oxidised to $CO_2$ and $H_2O$ via the Krebs cycle.

$\Delta G^o$
The Gibbs standard free-energy change ($\Delta G^{o'} = \Delta G^o$ at pH 7). Determined using 1.0 M of substrate when the reaction has gone to equilibrium. Measured in kJ mole$^{-1}$. A negative value indicates that energy is released during the reaction.

$\Delta G$
The actual free-energy change which occurs during the course of a reaction under physiological conditions. More useful than $\Delta G^o$, but more difficult to determine.

*α-helix*
The structure resulting when a polypeptide twists itself into a helical configuration and is stabilised by the formation of hydrogen bonds between the —CO residue of one peptide bond and an —NH residue three peptide bonds away.

*hexose*
A six-carbon (C6) sugar, e.g. glucose, fructose.

*hydrogen bond*
A bond arising from the attraction between the hydrogen of one group and the electrons surrounding the oxygen (or nitrogen) component of a second group. Hydrogen bonds have

|  |  |
|---|---|
|  | less than 1/10 of the strength of most biologically important bonds. |
| *hydrolysis* | Splitting with water. |
| *hydrophobic bond* | The tendency of hydrophobic groups to aggregate together. A slightly misleading term: bonds are not formed in the normal sense of the term. |
| *ion* | An atom with a positive charge (electron deficiency) or negative charge (electron excess). |
| *ionic bond* | A bond arising from the attraction which exists between oppositely charged ions. Ionic bonds are weak in solution because water molecules shield the charged groups from each other. |
| *isoenzyme* | An enzyme that occurs in different structural forms within a species or organism. The different forms may have slightly different properties. |
| *isomers* | Compounds with the same molecular formula but different spatial arrangements of the atoms. |
| *limiting factor* | Any factor, such as heat or light, which governs the rate of some process. |
| *negative feedback* | See feedback inhibition. |
| *nucleotide* | A compound derived from a C5 (pentose) sugar, a purine base (adenine, guanine) or a pyrimidine base (cytosine, thymine, uracil), and a phosphate. Constituents of ATP, NAD, NADP, FAD, RNA, DNA. |
| *oxidation* | The loss of an electron. In most biologically important reactions a proton normally accompanies the electron, and in such cases an oxidation is equivalent to a nett removal of hydrogen. Hence *oxidative respiration* involves the nett removal of hydrogen from carbohydrate ($CH_2O$) and its transfer to (say) |

oxygen, forming $H_2O$ (and $CO_2$). The loss of an electron by one compound is always accompanied by the gain of an electron by another, the latter reaction being called a *reduction*.

| *pacemaker* | The enzyme which determines the overall rate of flow through a biochemical pathway. |
|---|---|
| *pentose* | A five-carbon (C5) sugar, e.g. ribose, deoxyribose, ribulose. |
| *peptide bond* | The covalent bond formed by a condensation reaction between the carboxyl (—COOH) group of one amino acid and the amino (—$NH_2$) group of a second amino acid. |
| *pH* | An acidity–alkalinity scale over the range 0–14. Formally defined as $-\log_{10} [H^+]$. Since pure water has a hydrogen ion concentration of $10^{-7}$, the pH of pure water $= -\log_{10} (10^{-7}) = 7$. By corollary, in acids pH $< 7$, whilst in alkalis pH $> 7$. |
| *product* | A substance formed as a result of a chemical reaction. |
| *reactant* | A substance involved in a chemical reaction. |
| *reduction* | The gain of an electron (or hydrogen) cf. oxidation. |
| *secondary messenger* | A cytoplasmic substance, the presence or concentration of which is determined by a hormone (the primary messenger), and which causes a metabolic response characteristic of the hormone. |
| *substrate* | The substance consumed during a reaction. |
| *X-ray crystallography* | A technique which enables the three-dimensional structure of crystalline molecules to be deduced from the pattern produced when X-rays are passed through them. |

## GENERAL TEXTS

This volume was written to develop the themes outlined in general A-level text books, and specifically against the background given by the following:

*Advanced Biology* (2nd Edition), J. Simpkins and J. I. Williams, Bell and Hyman (1984).
*Biological Science*, Volumes 1 and 2, N.P.O. Green, G.W. Stout, D.J. Taylor, Cambridge University Press (1984, 1985).
*Biology* (4th Edition), Helena Curtis, Worth Pub. (1983).
*Biology: a Functional Approach* (3rd Edition), M.B.V. Roberts, Nelson (1982).
*Biology for Schools and Colleges*, Colin Clegg, Heinemann Educational Books (1980).

## SPECIFIC READING

### Enzymes, Energy (Chapters 1–3)

*Advanced Studies in Biology* Series (Basil Blackwell):
    *The Eukaryotic Cell*, M.R. Ingle (1985)
    *Genetic Mechanisms*, M.R. Ingle (1986).
*Carolina Biology Readers* (Carolina Biological Supply Co.):
    *ATP*, J.B. Chappell, No. 50 (1977)
    *The Mechanism of Enzyme Action*, M.R. Holloway, No. 45 (1976)
    *Hormones and Cell Metabolism*, 2nd Edition, P.J. Randle and R.M. Denton, No. 79 (1982).
*Studies in Biology* Series (Edward Arnold):
    *The Structure and Function of Enzymes*, 2nd Edition, C.H. Wynn, No. 42 (1979)
    *Temperature and Animal Life*, 2nd Edition, R.N. Hardy, No. 35 (1979)
    *Chloroplasts and Mitochondria*, 2nd Edition, M.A. Tribe and P.A. Whittaker (1982).

### Respiration, Autotrophy, Interconversions (Chapters 4–6)

*Carolina Biology Readers*:
    *Photorespiration*, A. Goldsworthy, No. 80 (1976)
    *The Energetics of Mitochondria*, 2nd Edition, J.B. Chappell,, No. 19 (1979).
*Studies in Biology Series*:
    *Bioenergetics of Autotrophs and Heterotrophs*, J.W. Anderson,, No. 126 (1980)
    *Photosynthesis*, 3rd Edition, D.A. Hall and K.K. Rao, No. 37 (1981)
    *The Biology of Respiration*, 2nd Edition, A. Bryant, No. 28 (1980).
*Photosynthetic Systems*, S.M. Danks, *et al.*, John Wiley (1983).

### Advanced Texts and References

The following titles may be of particular interest to more advanced students or their tutors:

*Biochemistry and Physiology of the Cell*, 2nd Edition, N.A. Edwards and K.A. Hassall, McGraw-Hill (1980).
*The Molecular Biology of the Cell*, B. Alberts *et al.*, Garland Press (1983).
*Outline Studies in Biology* (Chapman and Hall):
    *Control of Enzyme Activity*, 2nd Edition, P. Cohen (1983)
    *Enzyme Kinetics*, 2nd Edition, P.C. Engel (1981)
    *Hormone Action*, A.M. Malkinson (1975)
    *Metabolic regulation*, R.M. Denton and C.I. Pogson, 2nd Edition (1978).

Journals and Periodicals which may contain suitable articles include: *Biologist, J. Biol. Education, The School Science Review, New Scientist* and *Scientific American*.